COMPUTATIONAL NEUROSCIENCE:
Simulated Demyelinating
Neuropathies and Neuronopathies

COMPUTATIONAL NEUROSCIENCE:
Simulated Demyelinating Neuropathies and Neuronopathies

DIANA I. STEPHANOVA
Institute of Biophysics and Biomedical Engineering
Bulgarian Academy of Sciences, Sofia, Bulgaria

AND

BOZHIDAR DIMITROV
Institute for Population and Human Studies
Bulgarian Academy of Sciences, Sofia, Bulgaria

CRC Press
Taylor & Francis Group
Boca Raton London New York

CRC Press is an imprint of the
Taylor & Francis Group, an **informa** business

A SCIENCE PUBLISHERS BOOK

CRC Press
Taylor & Francis Group
6000 Broken Sound Parkway NW, Suite 300
Boca Raton, FL 33487-2742

© 2013 Copyright reserved
CRC Press is an imprint of Taylor & Francis Group, an Informa business

No claim to original U.S. Government works

Printed in the United States of America on acid-free paper

International Standard Book Number: 978-1-4665-7832-6 (Hardback)

Visit the Taylor & Francis Web site at
http://www.taylorandfrancis.com

CRC Press Web site at
http://www.crcpress.com

Science Publishers Web site at
http://www.scipub.net

Preface

In the field of computational neuroscience, this book is an attempt: (1) to summarize demyelinating neuropathies such as Charcot-Marie-Tooth Disease Type 1A (CMT1A), Chronic Inflammatory Demyelinating Polyneuropathy (CIDP), CIDP subtypes, Guillain-Barré Syndrome (GBS), Multifocal Motor Neuropathy (MMN) and neuronopathies such as Amyotrophic Lateral Sclerosis (ALS), simulated by our own models; (2) to compare the abnormalities in their axonal excitability properties; and (3) to explain the mechanisms underlying these abnormalities. The book, as a huge review, contains Introduction (Chapter I), Methods (Chapter II), Results, Discussions (Chapters III, IV) and References.

A brief description of the objects under the simulation and clinical studies of the nerve excitability properties, investigated by a threshold tracing technique in control groups and patients with CMT1A, CIDP, CIDP subtypes, GBS, MMN, ALS as well as a brief chronology of the models used for the simulation of nerve fibres are given in Chapter I.

Our multi-layered model used in computations is described in detail in Chapter II. Methods for both stimulation of human myelinated motor nerve axons and the calculation of their excitability properties such as potentials (action, electrotonic, extracellular), strength-duration time constants, rheobasic currents and recovery cycles are also given in this chapter.

Abnormalities in the multiple investigated axonal excitability properties and their mechanisms, for simulated normal and abnormal (CMT1A, CIDP, CIDP subtypes, GBS, MMN, ALS) cases are described, compared and discussed in detail in Chapter III. The complex structure of the myelin sheath is not taken into account in these simulations. The results confirm that the changes obtained in simulations replicate those recorded *in vivo* in control groups and patients with corresponding diseases. The results also confirm that axonal excitability properties are not identical, and they can be used as specific indicators for these disorders. The analysis shows that: (1) mild internodal systematic demyelination (ISD) is a specific indicator for CMT1A; (2) mild paranodal systematic demyelination (PSD) and paranodal internodal systematic demyelination (PISD) are specific indicators for CIDP and its subtypes; (3) severe focal demyelinations, each of them internodal and paranodal, paranodal-internodal (IFD and PFD, PIFD), are specific indicators for acquired demyelinating neuropathies such as GBS and MMN; (4) simulated progressively greater degrees of axonal dysfunctions termed ALS1, ALS2 and ALS3 are specific indicators for ALS Type1, Tape2 and Type3; and (5) the obtained excitability properties in simulated demyelinating neuropathies are quite different from those in simulated ALS subtypes, because of the different fibre electrogenesis. The results show that the abnormalities in the axonal excitability properties in the ALS1 subtype are near normal. The results also show that in simulated hereditary, chronic and acquired demyelinating neuropathies, the slowing of action potential propagation, based on myelin sheath dysfunctions, is larger than this, based on the progressively increased uniform axonal dysfunctions in simulated ALS2 and ALS3 subtypes. Conversely, abnormalities in the accommodative and adaptive processes are larger in the ALS2 and ALS3 subtypes than in demyelinating neuropathies. The increased axonal superexcitability in the ALS2 and ALS3 subtypes leads to repetitive discharges (action potential generation) in the nodal and internodal axolemma beneath the myelin sheath along the fibre length in response to applied long-duration subthreshold polarizing current stimuli (accommodative processes) and to applied long-duration suprathreshold depolarizing current stimuli (adaptive processes). Moreover, in the case of adaptation, the axonal superexcitability leads to blockage of each applied third (testing) stimulus in the recovery cycle of the ALS2 case and to blockage of each applied second (testing) stimulus in the recovery cycle of the ALS3 case. This is a result of repetitive firing caused by the preceding applied stimulus [i.e., by the applied first (conditioning) and second (testing) stimuli in the ALS3 and ALS2 cases, respectively]. The resulting superexcitability of the nodal and internodal axolemma beneath the myelin sheath, caused by repetitive discharges based on the continuous activation and reactivation of ionic (mainly "transient" Na^+) channels in these compartments can be regarded as a prelude to cell (neuron) death. And it is probably the reason for motor neuron degeneration in this disease.

The complex structure of the myelin sheath is taken into account in the simulations presented in Chapter IV. And the effect of the myelin sheath aqueous layers on the excitability properties in simulated hereditary and chronic demyelinating neuropathies are investigated and compared. The results show that aqueous layers have an additional effect on the simulated cases. All excitability properties, except for refractoriness and strength-duration time constants, worsen in simulated hereditary demyelinating neuropathies such as CMT1A and Dejerine-Sottas syndrome (DSS) when the myelin lamellae and their corresponding aqueous layers are reduced. Myelin sheath aqueous layers improve all axonal excitability properties in simulated CIDP. Aqueous layers do not modulate the axonal excitability properties in the simulated CIDP subtypes. This is because the reciprocally opposed effects of the aqueous layers on these properties are neutralized when demyelinations are heterogeneous.

This book should be of great interest to computational neuroscientists, neurologists, neurophysiologists, biophysicists, biologists, pharmacologists, lecturers and students in these fields.

Contents

Abbreviations

PNS	:	Peripheral Nervous System
CNS	:	Central Nervous System
NGF	:	Nerve Growth Factor
CMT	:	Charcot-Marie-Tooth Disease
CMT1A	:	Charcot-Marie-Tooth Disease Type 1A
DSS	:	Dejerine-Sottas Syndrome
CIDP	:	Chronic Inflammatory Demyelinating Polyneuropathy
GBS	:	Guillain-Barré Syndrome
MMN	:	Multifocal Motor Neuropathy
ALS	:	Amyotrophic Lateral Sclerosis
TCC	:	Terminal Cytoplasmic Cord
ISD	:	Internodal Systematic Demyelination
PSD	:	Paranodal Systematic Demyelination
PISD	:	Paranodal Internodal Systematic Demyelination
IFD	:	Internodal Focal Demyelination
PFD	:	Paranodal Focal Demyelination
PIFD	:	Paranodal Internodal Focal Demyelination

Chapter I

Nerve Fibres

Myelinated Axons

Studies during the last decades have focused on the intricate structure of myelinated axons, mainly by exploring the build-up, development, maturation and, eventual their degradation in genetically modified mice. Thanks to increasingly fine and revealing techniques in biochemistry, biophysics and micro-imaging we are now in possession of a completely new picture and thorough knowledge of many fine details for the structure of myelinated axons (Kirschner and Caspar 1972, Kirschner et al. 1984, Quarles et al. 2006, Zu Hörste and Nave 2006, Heredia et al. 2007, Douglas and Popko 2009, McGregor et al. 2010, Dučić et al. 2011). Since this book goes in a slightly different direction, we will not present the subject in great depth. Instead, we will review the general principles and gross outcome of accumulated knowledge for myelinated axons, diagnostics and differentiation between several forms of demyelination.

Axons of the peripheral nervous system (PNS) and the central nervous system (CNS) are highly specialized structures, and they are endowed with excitable membranes capable of electrogenesis. Myelin is built either of Schwann cells, surrounded by a basal lamina in the conducting axons of PNS, or of oligodendroglial cells, the so-called white matter, in the CNS. This structure is valid for both motor and sensory fibres (Peles and Salzer 2000). Myelination in the CNS has many similarities, but also points of difference, with respect to myelination in the PNS. CNS nerve fibres are not separated by connective tissue nor are they surrounded by cell cytoplasm, and specific glial nuclei are not obviously associated with particular myelinated axons. Unlike the peripheral nerve, where the sheath is surrounded by Schwann cell cytoplasm on the inside and outside, the cytoplasmic tongue in the CNS is restricted to a small portion of the sheath. This glial tongue is continuous with the plasma membrane of the oligodendroglial cell through slender processes. One oligodendrocyte can myelinate as many as 40 or more separate axons (Salzer 2003, Simons and Trotter 2007, Ndubaku and de Bellard 2008).

According to the organization of Schwann cells matured peripheral axons are classified either as unmyelinated or myelinated. In mammals (including humans), unmyelinated axons comprise approximately 75% of axons in cutaneous nerves and dorsal spinal roots, approximately 50% in muscle nerves (Ochoa 1976), and approximately 30% in ventral roots (Coggeshall et al. 1974, Risling and Hildebrand 1982). Unmyelinated axon diameters range from 0.1 to 2 µm (to approximately 3 µm in humans). The axons are more or less completely submerged in longitudinal troughs formed along the Schwann cell surface. In PNS myelinated nerve fibre, a single axon is associated with a train of longitudinally arranged Schwann cells.

In general, myelinated fibres are organized into several different structural parts (domains): an initial segment, deriving from the axonal hillock of the neuron; the axon, tightly covered by a myelin sheath (internode); a discontinuation with lack of cover (node of Ranvier); and two adjoining regions—paranodal (immediately adjoining the node; each region corresponding to 2% to 3% of the internode) and juxtaparanodal [further away alongside the sheath; stereotype internodal region (STIN) –95% of the internode] (Bhat et al. 2001, Bhat 2003, Nave and Salzar 2006, Birchmeier and Nave 2008, Nave 2010, Trapp and Kidd 2004). The length of a node ranges around 1 µm, and nodes are interposed several hundred µms apart. The internodal length, approximately from 200 to 2,000 µm, is correlated to axon size. The diameter of a peripheral myelinated axon is measured in the STIN region, and it is in the interval of 1 to 20 µm. The numerical relations between the axon diameter, myelin sheath thickness and internodal length are based on the studies of many authors (Yates et al. 1976, Berthold 1978, Arbuthnott et al. 1980, Friede et al. 1981, Friede and Bischhauser 1982, Berthold et al.

1983, Friede and Beuche 1985, Nilsson and Berthold 1988, Garbay et al. 2000, Nave and Trapp 2008, Thaxton and Bhat 2009).

The myelin sheath is a multilamelar membrane. It consists of repeating units of double bilayers separated by 3 to 4 nm thick aqueous layers that alternate between the cytoplasmic and extracellular faces of the cell membrane (Inouye and Kirschner 1988a). Dehydrated myelin is unusual in that it is composed of 75–80% lipid (Garbay et al. 2000, Quarles et al. 2006, Saher et al. 2011) and 20–25% protein by weight, compared with ~50% of most other cell membranes (Willians and Deber 1993). Multiple lipids make up the myelin sheath and each sheath, with its own distinct physical properties, contributes to the structure, adhesive stability, and possibly the pathogenesis of the myelin membrane. The asymmetric distribution of lipid composition on the cytoplasmic and extracellular faces likely also plays an important role (Inouye and Kirschner 1988b). Myelin basic protein constitutes 20–30% of total protein by weight and is located only between the two cytoplasmic faces, where it acts as an intermembrane adhesion protein (Roomi et al. 1978).

Schwann cells possess two distinct plasma membrane surfaces: an inner *adaxonal* membrane that makes contact with the axon and an outer *abaxonal* membrane that makes contact with the basal lamina. The beginning of the myelination process encompasses clustering of axonal *contactin-associated protein* (CASPR), glial *neurofascin* and *contactin* around the site of the nascent paranodal axo-glial junction (Rios et al. 2000). It is of note that close cell-to-cell contact with Schwann cell microvilli is crucial for the formation of voltage-gated Na^+ channel clustered at the node (Peles and Salzer 2000, Martini 2001, Sherman et al. 2005). Usually one Schwann cell entraps one internode (Jessen and Mirsky 2005). Neural cell adhesion molecules, *integrin* and the ligand *laminin*, together with Nerve Growth Factor (NGF), stimulate the Schwann cell. There exists an enormous enrichment of voltage-gated Na^+ channels at the nodes of Ranvier (about 1000–1500 μm^2 concentration), while alongside the internode their concentration is < 25 μm^2 (Waxman and Ritchie 1993, Salzer 1997, 2003, Waxman 1997, Pedraza et al. 2001, Salzer et al. 2008). First, the channels are generated. Then they form clusters around Schwann cells, and in the course of myelination, neighboring clusters fuse to form the Ranvier node (Ching et al. 1999, Rasband et al. 1998, 1999, Rasband 2011). Other nodal components include specific gangliosides and other glycoconjugates (Sheikh et al. 1999, Gong et al. 2002).

In contrast, oligodendrocytes' processes in the CNS release NGF and neurotrophic cytokines so that CNS myelination is fuelled by diffusible factors, called soluble signals. Oligodendrocytes surrounding neighboring axons synthesize myelin proteins under the influence of sphingosine-phosphate receptors, and the membrane assembly occurs at a distance. Then the nucleus of the paranode is formed by transport via microtubules of RNA granules and ribosomes (Sherman and Brophy 2005). The clusters of Na^+ channels induced by soluble oligodendrocyte factors delineate the nodes in the CNS and are spaced at a distance approximately 100 times the axon caliber (Kaplan et al. 1997). Also, not all fibres are necessarily myelinated. It is assumed that the minimum caliber for myelination is 1 μm (Sherman et al. 2005). That is, tiny, predominantly sensory conducting fibres are normally non-myelinated; however they are frequently embedded within common ensheathment alongside streaks of myelinated fibers.

Although the final outcome of initial segments' structure—Na^+ channels included—closely resembles that of the nodes, the presence of glial cells is not obligatory for their formation. Another protein—*ankyrin G*, and intrinsic determinants from the cell, are supposed to establish a lateral diffusion barrier for integral membrane proteins (Winckler et al. 1999). Thus other cytoskeletal elements and cell adhesion molecules (such as NF186, NrCAM, βIV-Spectrin, etc.) prevail in the axonal initial segment area (Dzhashiashvili et al. 2007, Salzer et al. 2008, Rasband 2011).

The outermost part of the insulation—basal lamina—covers the compact myelin sheath alongside the internode. On either side of the node numerous Schwann cell microvilli project from the outer collar of the compact myelin. These lamellae open up into a series of paranodal cytoplasmic loops (PCL) that spiral around, closely appose and form a series of septate-like junctions with the axon. A major building component is the *contactin-associated protein* (CASPR), together with *glycosylphosphatidyl-inositol* (GPI) and *neurofascin* (Rios et al. 2000). They represent a potential site for bidirectional signaling between axons and myelinating glial cells. Cells expressing *proteoglycan* make contact with the nodal axolemma in CNS white matter (Butt et al. 1999). The axon is enormously enriched in intramembranous particles at the node that correspond to Na$^+$ channels. Na$^+$ channels in the node are anchored to the cytoplasmatic tail of proteins and ligands (*tenascin, neurofascin* etc.) to either the Schwann cells' microvilli or perinodal astrocytes. They also form macromolecular complexes with *ankyrin* G. Similar protein ligand bonds were described for the paranodal and juxtaparanodal sections.

Paranodes are noted by axonal periglial loops of transverse bands that physically invaginate the axolema. Adhesion between the axon and glia is such that a firm barrier is built against diffusion of ions from the node towards the internode in the process of generation of the action potential (Rios et al. 2003). Internodes, covered by the inner membrane of Schwann cells, provide a periaxonal space above the underlying axon of about 15 nm. This space is rigorously maintained throughout development, and in many pathologic states, but might be disrupted by the action of some proteases (Salzer et al. 2008).

The Schwann cell per axon unit is surrounded by a basement membrane, built of protein *laminin* and binded by receptors like *periaxin* and *dystroglycan complexes*. These are clustered between the *abaxonal surface* of the myelin sheath and the Schwann cell plasma membrane, thus forming cytoplasmic channels. This was described as *longitudinal bands* by Ramón y Cajal and named after him. They regulate the increase in length of Schwann cells in the course of the organism's growth (Sherman and Brophy 2005).

The region of the axon that lies under the myelin sheath beyond the innermost paranodal junction (juxtaparanodal section) contains concentrated potassium channels and *contactin-associated protein-2*. The K$^+$ channels promote repolarization and maintain the internodal resting potential. K$^+$ channel clustering occurs well after the formation of Na$^+$ channels, and therefore the juxtaparanodal segregation depends on the previous paranodal formation. It was experimentally proven that nodes are able to form independently from the paranodal junctions, whereas the junctions appear to be a prerequisite for the delineation of axonal domains, including the appropriate localization of K$^+$ channels to the juxtaparanodes (Dupree et al. 1999).

Barrett and Barrett (1982) have presented evidence that the internodal axolemma and its strongly rectifying channels contribute to post-spike changes in membrane potential and excitability, and a role for these channels in determining resting potential is argued by many authors (Rithchie and Chiu 1981, Chiu and Ritchie 1984). Baker et al. (1987) apply long-duration (100 ms) current stimuli to intact rat spinal roots treated with tetrodotoxin to block sodium currents, and they recorded fast (rise time < 1 ms) and slow (10 100 ms) components of the non-propagated changes in membrane potential (electrotonus), as first described by Lorente de Nó (1947). Pharmacological dissection of the components of electrotonus indicates that at least three types of internodal ion channels contribute to the membrane potential of intact fibres: fast potassium channels (K_f), sensitive to 4-aminopyridine (4AP); slow potassium channels (K_s), sensitive to tetraethylammonium (TEA); and inwardly rectifying channels (IR), sensitive to caesium ions (Baker et al. 1987). These channels were also investigated by many other authors (Roper and Schwarz 1989, Scholz et al. 1993, Safronov et al. 1993, Vogel and Schwarz 1995, Zhou et al. 1998).

Action potentials are generated at the axon hillock in the initial segment area, propagate to and are further refreshed at the periodic interruptions between myelin segments—Ranvier nodes. Current flows inward at the node, reflecting the high local concentration of voltage-gated Na^+ channels, and propagates to the next node of Ranvier because of the high resistance and low capacitance of the surrounding myelin sheath. This is known as saltatory conduction type. The extent of myelination together with the length of the internodes, determine conduction velocity and provide a mechanism for synchronization of presynaptic input to afferent targets (Waxman 1997). The loss of the Na^+ channels leads to decreased conduction velocity and to disturbance of the nodal and paranodal structure (Kearney et al. 2002).

The architecture of myelinated axons possesses the features of a so-called "double-edged sword", according to one recent comment (Nave 2010). On one hand, the glial sheath reduces energy for neuronal communication and boosts the speed of impulse propagation via the mechanism of saltatory excitation. On the other hand, however, it restricts access to extracellular metabolic substrates, creating hazards for axonal vitality. It was established that glycolytic metabolism prevails over oxidative metabolism in the white matter (Morland et al. 2007). Glycolytic enzymes synthesized in the soma of nerve cells move anterogradely by the rate of 2 mm/daily. This is such a slow axonal transport rate that distal internodes in myelinated axons of, say, 1 meter, would only receive them after a hypothetical period of 500 days (!). This is provided that they are not already otherwise totally degraded by lysosomes (the glucose-6-phosphate dehydrogenase half-life in erythrocytes is just 50 days...). Nature has cleverly insured against such pitfalls by introducing a thin cylindrical, liquid-filled periaxonal space that separates those axons containing microtubules from those containing mitochondria. This channel-like system forms an inner collar of the myelin sheath, and it is interrupted only by the Schmidt-Lanterman incisures, which spiral it around the axon similar to paranodal loops. Both loops and incisures provide a direct route from the glial soma to axons. Therefore, the fast axonal transport rate—of 2–5 μm/s—is guaranteed by the intracellular convection system with molecular motors and tubulin tracts (Lasek et al. 1977, 1984, Court et al. 2008). In the CNS the mechanism is again somewhat different. Oligodendrocytes make contact with astrocytes, and they in their turn directly tap into the small capillaries. Adjacent membranes of axon and oligodendrocytes form gap junctions, and the exchange of metabolites is in both directions (Moffett et al. 2007).

Demyelinating Neuropathies

Disordered domain organization and subsequent axonal degeneration initially disrupt the velocity of signal transmission. This is followed by conduction block in severe cases. There are several factors leading to such disturbances: inherited faulty architectural build-up; disorders acquired in the course of life (trauma, inflammation, intoxication); and wear out due to age-related deficits. Common demyelinating diseases of the peripheral nerve include: Charcot-Marie-Tooth Disease, Guillain-Barré Syndrome, Multifocal Motor Neuropathy, Chronic Inflammatory Demyelinating Polyneuropathy, Multiple Sclerosis, etc. Disintegration of the domain organization of the myelin nerve fibre, sometimes leading to an axonal degeneration, is a characteristic feature of these diseases (Kimura 2001, Lubetzki et al. 2005, Oguivetskaia et al. 2005, Shy 2006, Berger et al. 2006, Nave et al. 2007, Trapp and Nave 2008). Some cases of diabetic and carcinomatous neuropathies also belong to this category, although most paraneoplastic syndromes show axonal degeneration rather than demyelination.

Charcot-Marie-Tooth Diseases (CMT) and Type IA (CMTIA)

Hereditary neuropathies of the PNS are a clinically heterogeneous group of disorders. Recent advances in genetic research have identified various types of CMT (CMT1A, CMT1B, CMT1C, CMT2, CMT3, CMT4, CMTX1, CMTX2), and those correspond to specific genetic mutations (Dyck et al. 1993, Kimura 2001, Young and Suter 2001, 2003, Shy et al. 2002, Suter and Scherer 2003). About 70% of CMT are autosomal dominant, however, recessive and X-linked varieties also exist. The major types are: Charcot-Marie-Tooth disease Type1 (CMT1) or hereditary motor and sensory neuropathy Type I (HMSN I), a demyelinating neuropathy; Charcot-Marie-Tooth disease Type2 (CMT2) or hereditary motor and sensory neuropathy Type II (HMSN II), an axonal neuropathy; and Dejerine-Sottas syndrome (DSS) or hereditary motor and sensory neuropathy Type III (HMSN III), a severe demyelinating neuropathy. CMT Type1 is a disorder of peripheral demyelination resulting from a mutation in the peripheral myelin protein (PMP) 22 gene. This mutation results in abnormal myelin, which is unstable and spontaneously breaks down. This causes segmental demyelinations along the fibre length, which in turn slows the conduction velocity in the motor and sensory nerves (Henriksen 1956, Gilliatt and Thomas 1957, Dyck and Lambert 1968a,b, 1974, Dyck et al. 1993, Harding and Thomas 1980, Ohnishi et al. 1987, Baraister 1990, Kimura 2001, Fabrizi et al. 2007, Hattori et al. 2003, Verhamme et al. 2009, Yiu et al. 2008). CMT Type2 reveals axonal loss with Wallerian degeneration, but the conduction velocities of the propagated impulses through the CMT2 axons are near normal (Dyck et al. 1993, Guertin et al. 2005, Moldovan et al. 2009, Coleman and Freeman 2010). CMT Type3 results in severe demyelination and is much more severe than CMT Type1. Dejerine-Sottas syndrome (CMT3) is characterized by considerable slowing (2–5 m/s) or blocking of conduction in motor and sensory nerves (Benstead et al. 1990, Dyck et al. 1993, Gabreëls-Festen 2002, Yim et al. 1995). Charcot-Marie-Tooth disease Type 1A (CMT1A) is an autosomal dominant demyelinating polyneuropathy usually associated with a large DNA duplication of chromosome 17. CMT1A has a tandem duplication of chromosome 17p11.2–12 with trisomic expression of the peripheral myelin protein 22 (PMP-22) gene or, less frequently, a missense mutation of PMP-22. This disorder is the most common form of hereditary neuropathy. Its hallmark is uniform demyelination, which results in slowing of conduction velocity in the motor and sensory nerves (Benstead et al. 1990, Lupski et al. 1991, Raeymaekers et al. 1991, Dyck et al. 1993, Birouk et al. 1997, Yiu et al. 2008, Verhamme et al. 2009). Because CMT is an inherited disorder, patients almost always have a family history of the disease. There are, however, rare sporadic cases without family history, and these likely represent spontaneous mutations. CMT is characterized by slowly progressive weakness and numbness that begins in the distal limb muscles, and this typically occers in the legs before the arms. Symptoms usually appear in the first two decades of life (Marques et al. 2005, Houlden and Reilly 2006). Currently, there is almost no treatment for the underlying disorder, nothing which can fix the abnormal myelin or prevent its degeneration. Sahenk et al. (2005) suggest neurotrophin-3 (NT3), which promotes nerve regeneration and sensory improvement in CMT1A mouse models and patients, whereas Martini and Toyka (2004) discuss immune-mediated components in CMT1A mouse models and patients. Improving understanding of the genetics and biochemistry of this disorder, however, offers hope for an eventual treatment.

Chronic Inflammatory Demyelinating Polyneuropathy (CIDP) and its Subtypes

Chronic inflammatory demyelinating polyneuropathy (CIDP) as a peripheral nerve disorder is one of several chronic demyelinating neuropathies that are believed to have autoimmune etiology in which the immune system aberrantly recognizes one or more molecular components of the myelin sheath as a foreign object; resultant antigen per antibody reactions are believed to underlie disruptions of the structural integrity of the sheath (Rezania et al. 2004). There is no evidence of any genetic basis and inheritance of this disease. CIDP is a disorder with diffuse demyelination, which results in slowing or blocking of action potential conduction in motor and sensory nerves (Köller et al. 2005). This disorder is also characterized by complete symmetry, as both proximal and distal muscles are affected (Köller et al. 2005, Saperstein et al. 2001). CIDP can occur concurrently with other systemic diseases (Barohn et al. 1989, Gorson et al. 2000, Katz et al. 2000, Köller et al. 2005), and there are subtypes of chronic demyelinating neuropathies (Lewis et al. 1982, Krarup et al. 1990, Katz et al. 1997, 2000, Christova et al. 2001, Alaedini et al. 2003, Alexandrov et al. 2003) that are broadly classified under the term of chronic inflammatory demyelinating polyneuropathy. The disease may follow a progressive course over several years with severe generalized disability, or it may only affect a upper limb. Prednisone causes a small but statistically significant improvement over no treatment (Dyck et al. 1982, Feasby 1996). Plasma exchange is a useful therapy, especially in cases with features of demyelination rather than axonal degeneration (Pollard et al. 1983, Gorson et al. 1997). Additional modes of therapy include cyclosporine (Hodgkinson et al. 1990, Barnett et al. 1998), immunoglobulin (van Doorn et al. 1991, Tan et al. 1993, Vermeulen et al. 1993, Dyck et al. 1994, Hiroso et al. 1996, Hughes 2002, Eftimov et al. 2009, Boërio et al. 2010) and interferon—α (Gabatelli et al. 1995, Gorson et al. 1998).

Guillain-Barré Syndrome (GBS) and Multifocal Motor Neuropathy (MMN)

Although of unknown etiology, GBS and relative demyelinating neuropathies closely resemble experimental allergic neuritis (Schmidt et al. 1992). Some patients with this syndrome have human immunodeficiency virus (HIV), herpes zoster virus or hepatitis B virus infections. In most patients, however, repeated attempts have failed to isolate infective agents. These findings support an autoimmune pathogenesis rather than direct invasion of the nerve by infections agents (Hartung et al. 1995). Segmental conduction slowing or block in certain parts of the nerve characterizes acquired inflammatory demyelinating neuropathies with nonuniform involvement. The segment of maximal involvement varies from one patient to the next. This helps to explain the diversity of clinical findings and of conduction abnormalities in different cases.

GBS consists of a number of subtypes showing different clinical and pathological features. GBS is classified by pathological and electrodiagnostic criteria (Feasby et al. 1986, Griffin et al. 1995, Choudhury and Arora 2001) into the following subtypes: acute inflammatory demyelinating polyneuropathy (AIDP); Fisher syndrome—acute motor and sensory axonal neuropathy (AMSAN); and acute motor axonal neuropathy (AMAN). After treatment, conduction studies may or may not revert toward normal values (Toyka et al. 1980, Vriesendorp et al. 1991).

According to some authors, MMN is a unique variant of CIDP (Parry 1985, Parry and Clarke 1985, Pestronk et al. 1988). A critical diagnostic feature in MMN is the demonstration of conduction block in the multiple peripheral nerves (van der Bergh et al. 1989, Krarup et al. 1990, Story and Phillips 1995, Sumner 1997, Kimura 2001, Kaji 2003, Priori et al. 2005). The long-duration conduction block suggests chronic demyelination as the pathological basis. Both motor conduction block and increased threshold probably reflect a chronic focal demyelinating lesion that, for yet undetermined reasons, becomes persistent without repair (Kaji et al. 1994, Yokota et al. 1996, Kimura 1997). Several authors have documented successful treatment with intravenous immunoglobulin (Krarup et al. 1990, Kaji et al. 1992, Parry 1997, van der Bergh et al. 1998, Boërio et al. 2010). Some patients improve but do not return to normal, others stabilize, some require long-term therapy, and still others become refractory to any form of treatment. Most studies suggest better results with cyclophosphamide or human immunoglobulin therapy (Chaudhry et al. 1993, Noble-Orazio et al. 1993) than with prednisone or plasmapheresis.

Neuronopathies

Amyotrophic Lateral Sclerosis (ALS)

ALS, also known in the USA as Lou Gehrig's disease, is the most common of the motor neuron diseases. It is a neurodegenerative, and usually fatal, disorder involving the neurons and the motor system pathways in the brain and spinal cord. ALS is characterized by progressive muscle atrophy starting in the limbs and spreading to the rest of the body, often accompanied by overactive reflexes.

The median age of onset is 55–66 for the sporadic type of ALS. Onset before age 20 is rare, but it does occur. ALS may be fatal in one year or continue for 10 or more years. Familial ALS is genetically determined. It is inherited as an autosomal dominant trait. Since the discovery that some patients with familial ALS show a mutation in the superoxide dismutase (SOD1) gene on chromosome 21q (Rosen et al. 1993), further progress has been made in the understanding of the pathogenesis of ALS (Brown 1997, Cleveland 1999). In most instances, sporadic and autosomal dominant familial ALS are clinically similar. The most consistent features in ALS are repetitive discharges (Trojaborg and Buchthal 1965, Wettstein 1979, Conradi et al. 1982, Roth 1982, 1984, Sobue et al. 1983, Finsterer et al. 1997, Rowinska-Marcinska et al. 1997, Daube 2000) and increased axonal excitability (Trojaborg and Buchthal 1965, Guiloff and Modarres-Sadeghi 1992). Another ALS feature is a minor slowing of motor nerve conduction (Sobue et al. 1983, Nakanishi et al. 1989, Wirguin et al. 1992, Arasaki and Tamaki 1998, Feinberg et al. 1999). Riluzole (Rilutek TM) is the first and only drug to receive approval for the treatment of ALS. This drug has shown a modest prolongation in patient survival time (Bensimon et al. 1994, Lacomblez et al. 1996). Studies for other drugs have conflicting results or are not yet completed (Dittrich et al. 1996, Lai et al. 1997, Miller and Sufit 1997, Iwasaki et al. 1998, Ludolph et al. 1999).

Axonal Excitability

Axonal excitability problems of the nerve-muscle system are still of current interest since our entire body activity is based on the activation and interaction between neurons, nerves and muscles. It is well known that an accurate diagnosis of a given disorder depends in part on the application of the appropriate electrodiagnostic technique and on electrophysiological studies. The nerve conduction study, which is part of the laboratory studies that include blood tests, muscle or nerve biopsy, genetic testing, etc. is still successfully realized. The major function of nerve axons is to transmit information reliably from one site to another. If the axon of a myelinated fibre remains in continuity despite alterations or demyelinations, the fibre may retain its ability to conduct nerve impulses, but the axonal dysfunctions and demyelinations lead to a spectrum of conduction abnormalities. Conduction slowing or conduction block is a characteristic feature already discussed for CMT1A, CIDP (Cappelen-Smith et al. 2001, Nodera et al. 2004, Sung et al. 2004), n-hexane neuropathy (Chang et al. 1998, Kuwabara et al. 1993), GBS and MMN (Kaji 2003, Kuwabara et al. 2002, Priori et al. 2005). Action potential conduction velocities are near normal in ALS subtypes (Sobue et al. 1983, Nakanishi et al. 1989, Wirguin et al. 1992, Arasaki and Tamaki 1998, Feinberg et al. 1999) and consequently, they cannot be regarded as definitive indicators for the progressive degrees of this disease.

In the 1990s, a non-invasive threshold tracking technique was developed (Bostock and Baker 1988, Bostock et al. 1998, Kiernan et al. 2000) to measure "threshold electrotonus", which revives methods of studying nerve accommodation and measures excitability indices (such as strength-duration time constants, rheobasic currents and recovery cycle) in control groups and patients with CMT1A, CIDP, GBS, MMN, ALS (Cappelen-Smith et al. 2001, Kaji 2003, Kuwabara et al. 2002, 2003, Nodera et al. 2004, Nodera and Kaji 2006, Priori et al. 2005, Sung et al. 2004). These electrophysiological studies show very similar depolarizing responses; however, they show different hyperpolarizing responses of threshold measurements in patients with CMT1A, CIDP and CIDP subtypes. Their threshold electrotonus changes are greater in response to the hyperpolarizing stimuli (Cappelen-Smit et al. 2001, Nodera et al. 2004, Sung et al. 2004) than in the normal controls. Threshold electrotonus, in response to depolarizing or hyperpolarizing current stimuli, is slightly changed in GBS and MMN (Kuwabara et al. 2002, Kiernen et al. 2002); whereas it is abnormal with spontaneous discharges in ALS subtypes (Bostock et al. 2005). Using the threshold tracking technique, it was found by Bostock et al. (2005) that while some responses are normal, others show decreases or rapid increases in threshold to 40% polarizing current stimuli. The abnormal responses to 40% polarizing current stimuli observed in ALS patients ware classified as Type1, Type2 and Type3 (Bostock et al. 1995). It was supposed that these abnormal responses are associated with progressively increased axonal potassium ion channel dysfunction. Impaired potassium (K^+) channels in ALS patients were suggested by Bostock et al. (1995). Using the same technique, clinical investigations of the above mentioned diseases show that axonal excitability properties such as strength-duration time constants, rheobases and recovery cycles are also abnormal. These include longer strength-duration time constants in patients with CMT1A and ALS or shorter strength-duration time constants in patients with CIDP, GBS and MMN than in normal controls. It was also shown in clinical studies that the less refractoriness in the relative refractory period of the recovery cycle is a characteristic feature in patients with CIDP, GBS, MMN and ALS; whereas it is near normal in CMT1A in comparison with normal controls. Greater superexcitabilities in the recovery cycles are observed for all investigated demyelinating neuropathies and ALS Type1. However, abnormal superexcitabilities of the axons are observed in the recovery cycles for ALS Type2 and Type3.

We have simulated the same disorders (Daskalova and Stephanova 2001, Stephanova and Daskalova 2002, 2005a,b, 2008, Stephanova and Alexandrov 2006, Stephanova et al. 2005, 2006a,b, 2007a,b, 2011a,b, Krustev et al. 2010), to reveal the mechanisms of abnormalities found in their axonal excitability properties. In Chapter III: (1) existing simulations of the above mentioned demyelinating neuropathies and ALS subtypes are summarized; (2) abnormalities in their multiple investigated axonal excitability properties are shown; and (3) explanations of mechanisms underlying these abnormalities are given and discussed.

Mathematical Modeling of Nerve Fibres

Mathematical models are developed to examine the possible influence of anatomical and electrical properties on the functional behavior of excitable structures such as muscle and nerve (unmyelinated and myelinated) fibres. The first computational model, based on the ion theory of nerve impulse propagation through the nodal membrane, was developed by Hodgkin-Huxley-Katz (Hodgkin et al. 1949, 1952, Hodgkin and Huxley 1952a–d).

Modified Hodgkin-Huxley models have been subsequently used successfully to simulate not branching or branched unmyelinated axons. The study by Moradmand and Goldfinger (1995) considers if the coding of complex random impulse trains is distorted by long-distance propagation in unmyelinated axons. They found that short interval sequences are modified due to impulse attenuation in the wake of a recent preceding impulse, as assessed with stochastic steady-state and real-time estimators. Investigations of the action potential propagation at a bifurcation of unmyelinated axons showed that the branch points are a highly-efficient substrate for impulse divergence (Goldfinger 2000). In this study the passive properties and boundary conditions at a branch point are determined using analog circuitry to reconstruct solutions to the cable equation. It was found that a symmetrical branch point is a linear divider of axial current. This enables the exact cable representation of the boundary conditions at any branch point, and disproves Rall's assumed boundary conditions. New data about the theory of impulse propagation at axonal branch points in unmyelinated axons are provided by Goldfinger (2005a,b). Among the results are: (1) impulse propagation across secondary and tertiary branch points has a wide-band efficiency; (2) K^+ accumulations in the periaxonal space modify impulse shape depending upon the spatial extent of such accumulations and attenuated short interspike interval generation; (3) Markovian kinetics for Na^+ channel transitions enhances the velocity of impulse propagation (Goldfinger 2005a); and (4) Rall's theory of equivalence (i.e., that dendritic or axonal branching could be represented by an 'equivalent cylinder') is reexamined. It was shown that literal branching has passive and active cable properties entirely different from their supposed 'equivalent' (Goldfinger 2005b). Goldfinger (2009) uses the voltage-dependent Markovian model to analyze the molecular basis for relative refractoriness in unmyelinated axons. The findings include: (1) at rest, the probability distributions for Na^+ channels in a given transition state are not equal, and include a quasi-basin of stability in one of the four inactive states; (2) supernormality in the relative refractory state is attributed to the relative speed of transitions out of the closed states.

The classical model of myelinated fibres is based on the Hodgkin-Huxley theory with a nodal membrane described by Fitzhugh (1962) in which the myelin sheath is collapsed. Goldman and Albus (1968) modified this model to include a description of the nodal membrane derived from the experimental data on Xonopus Leavis myelinated nerve fibres as described by Frankenhaeuser and Huxley (1964) [Dodge and Frankenhaeuser 1959, Frankenhaeuser and Waltman 1959]. These models assume low resistance nodes and a high resistance myelin sheath which is an effective insulator of the internodal axolemma. Most subsequent models of myelinated fibres have followed this single cable formulation (e.g., Smith and Koles 1970, Koles and Rasminsky 1972, Schauf and Davis 1974, Waxman 1977, 1978, Waxman and Brill 1978, Waxman et al. 1979, Wood and Waxman 1982, Wood et al. 1982, Kocsis et al. 1982, Waxman and Wood 1984, Stephanova 1988a,b,c, 1989a,b, 1990, Quandt and Davis 1992). In contrast, Blight (1985) represents the internodes as double cables, with separate potential gradients occurring across the internodal axolemma and across the myelin sheath, which is assumed to provide a relatively low-resistance pathway compared to the classical models. His model describes the contribution of the passive discharge of the capacitance of the internodal axolemma to intracellular recorded depolarizing afterpotentials (Barrett and Barrett 1982, Blight and Someya 1985). Halter and Clark (1991) have enlarged on the Blight model, by considering the consequences of the complex geometry of the nodal and paranodal regions. Other models of myelin axons have taken into account evidence for the non-uniform distribution of ion channels along the internodal axolemma and additional channel types to those considered in the classical models (Chiu and Ritchie 1981, 1984, Brismar and Schwarz 1985, Chiu and Schwarz 1987, Baker et al. 1987, Roper and Schwarz 1989, Chiu et al. 1999). Bostock et al. (1991) use a simple two-compartment (node + internode) model to test whether Stampfli's (1959) proposal for two stable states of axons in high potassium could be applied to post-ischaemic human nerve. Their model uses nodal and internodal channel types previously demonstrated in myelinated axons (Baker et al. 1987, Chiu and Schwarz 1987, Bostock and Baker 1988, Shrager 1989) with channel permeabilities adjusted to match the responses of the model and human axons to polarizing current stimuli. Our double cable model of human motor nerve axons with a long chain of nodes, paranodes and double cable internodes (Stephanova and Bostock 1995, 1996) is an extended version of the Bostock model et al. (1991). Unlike Blight (1985), we have assumed a high-resistance myelin sheath (as measured by Tasaki 1955), and a leakage pathway via the paranodal junction. The functional behavior of myelinated nerve axons was recently investigated using double cable models (Dimitrov 2000, 2005, 2009, Gow and Devaux 2008, Schiefer 2009, Zlochiver 2010).

We have also proposed a new multi-layered model (Stephanova 2001) in which the complex structure of the myelin lamellae are taken into account. In this model, which is still the first and only in the literature, the myelin aqueous layers provide appreciable longitudinal and radial conductance. We investigated the contribution of these conductive paths to the electrical properties of human motor nerve axons, representing the myelin sheath as a series of interconnecting parallel lamellae (Stephanova 2001).

The complex structure of the paranodal region, nodal resistive gap and "persistent" Na^+ channels are not taken into account in our multi-layered (Stephanova 2001) and double cable (Stephanova and Bostock 1995, 1996) models. The classical "transient" Na^+ currents are set at 100% in our simulations, whereas they are 99% in human motor nerve axons (Bostock and Rothwell 1997). The remaining 1% is reserved for the "persistent" Na^+ channels (Bostock and Rothwell 1997). Evidence for a low-threshold, "persistent" Na^+ current in human peripheral nerve has come from an analysis of the different excitability properties of motor and sensory axons, as shown by their strength-duration time constants measured by a method of latent addition (Bostock and Rothwell 1997). The authors were able to model their experimental data by replacing a small fraction of "transient" Na^+ channels (more at sensory (2%) than at motor (1%) nodes) with "persistent" Na^+ channels, activating at 20 mV more negative than the "transient" Na^+ channels and at half the rate.

Our multi-layered and double cable models are described in detail in Chapter II.

Models and Methods for Investigation of the Human Motor Nerve Fibre

Multi-Layered and Double Cable Models

The method of mathematical modeling and computational simulations of human motor nerve fibres in normal and abnormal cases are generally used to reveal some of the mechanisms underlying nerve excitability abnormalities observed and recorded in patients with demyelinating neuropathies and neuronopathies. The simulations presented here apply our model (Stephanova 2001), in which 150 interconnected parallel lamellae are simulated by alternating 150 lipid and 150 aqueous layers within the myelin sheath. The aqueous layers provide appreciable longitudinal and radial conductance, the latter via a spiral pathway (Fig. 1). This multi-layered model of the human motor nerve fibre is a further development of the double cable models (Blight 1985, Halter and Clark 1991, Stephanova and Bostock 1995, 1996). It is derived from the model of Stephanova and Bostock (1995) in which the myelin sheath and internodal axolemma are treated as two concentric cables (Fig. 2). All calculations are carried out for fibres with: an axon diameter of 12.5 μm; an external fibre diameter of 17.3 μm; nodal diameter of 5 μm; nodal area of 24 μm^2; myelin thickness of 2.4 μm; periodicity of myelin lamellae of 16 nm and periaxonal space thickness of 20 nm. The temperature is 37°C.

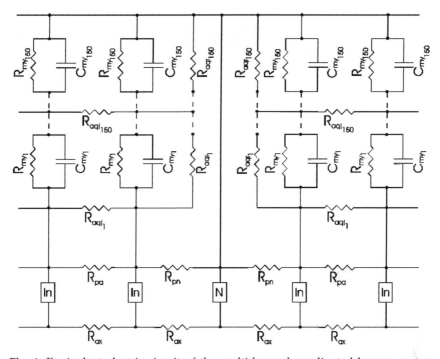

Fig. 1. Equivalent electric circuit of the multi-layered myelinated human motor nerve fibre (from Stephanova 2001). One node (N box) and internodal segments, each including internodal axolemma (In boxes) and the multi-layered myelin sheath are presented. The longitudinal axoplasmic (R_{ax}), periaxonal (R_{pa}) and paranodal seal (R_{pn}) resistances are also illustrated. The equivalent electric circuit for the multi-layered myelin sheath contains, respectively: myelin-layered capacitance and resistance (C_{myN}, R_{myN} for N = 1,150), as well as aqueous-layered longitudinal and radial resistances (R_{aqlN}, R_{aqrN} for N = 1,150). The equivalent circuits for each of the N and In boxes are given in the *bottom row* of Fig. 2, representing the double cable model (Stephanova and Bostock 1995) of human motor nerve fibre.

From the electric equivalent circuits (Figs. 1 and 2) the following set of partial differential equations is derived from Kirchoff's current law:

$$C_a(x)\,\frac{\partial V_a(x, t)}{\partial t} = I_s(t) + \frac{1}{R_{ax}}\,\frac{\partial^2 V(x, t)}{\partial x^2} - q_i(x, t) \tag{1}$$

$$C_{my}\,\frac{\partial V_m(x, t)}{\partial t} = \frac{1}{R_{ax}}\,\frac{\partial^2 V(x, t)}{\partial x^2} + \frac{1}{R_{pa}(x)}\,\frac{\partial^2 V_m(x, t)}{\partial x^2} - \frac{V_m(x, t)}{R_{my}} \tag{2}$$

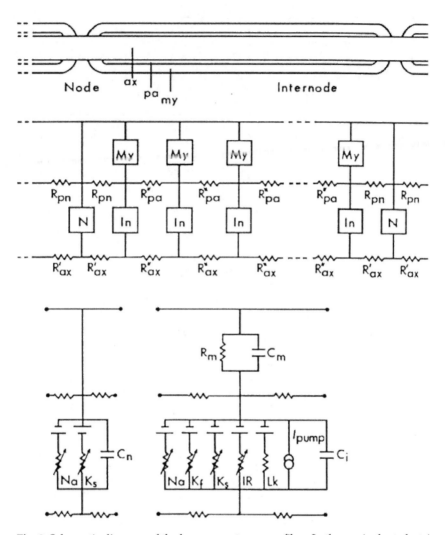

Fig. 2. Schematic diagram of the human motor nerve fibre. In the equivalent electric circuit, two consecutive nodes (N boxes), adjusted distal paranodes and internodal segments between them, each including myelin sheath (My boxes) and internodal axolemma (In boxes), are schematically presented. A paranodal seal resistance (R_{pn}) joins each internode to the adjacent nodes of Ranvier. In accordance with the non-uniform spatial step sizes, the various longitudinal axoplasmic (R_{ax}) and periaxonal (R_{pa}) resistances are illustrated by their printed elements. In the *bottom row*, the equivalent electric circuits for each of the boxes contain, respectively: channels (Na, K_s) and capacitance (C_n) for N boxes; channels (Na, K_f, K_s, IR, Lk), electrogenic pump (I_{pump}) and capacitance (C_i) for the In boxes: and resistance, capacitance (R_m, C_m) for My boxes. Channels: Na (sodium), K_f (fast potassium), K_s (slow potassium), IR (inward rectifier), Lk (leak), (from Stephanova and Bostock 1995).

$$C_{my1}\frac{\partial V_1}{\partial t} + \frac{V_1}{R_{my1(x)}} = \frac{\partial^2 V}{\partial x^2}\frac{1}{R_{ax}} + \frac{\partial^2 V_1}{\partial x^2}\left(\frac{1}{R_{aql1}} + \frac{1}{R_{pa}}\right) = I_1 \tag{3}$$

$$C_{my2}\frac{\partial V_2}{\partial t} + \frac{V_2}{R_{my2(x)}} = I_1 + \frac{\partial^2 V_2}{\partial x^2}\frac{1}{R_{aql2}} = I_2 \tag{4}$$

$$\vdots$$

$$C_{myN}\frac{\partial V_N}{\partial t} + \frac{V_N}{R_{myN(x)}} = I_{N-1} + \frac{\partial^2 V_N}{\partial x^2}\frac{1}{R_{aqlN}} = I_N \tag{5}$$

In the paranodal space

$$R_{myN(x)} = 1/(\frac{1}{R_{myN}} + \frac{1}{R_{aqrN}}) \text{ for } N=1,150 \tag{6}$$

where R_{aqr150} is infinite and in the other parts of the internodal space

$$R_{myN(x)} = R_{myN} \text{ for } N=1,150 \tag{7}$$

This set of partial differential equations describes the potential (V_a) and ionic currents [$q_i(x, t)$] either across the nodal or across the internodal axolemma, as well as potentials and currents, respectively, across the myelin (V_m, I_m) and its N-layers (V_N, I_N), for N=1, 150. The ionic currents $q_i(x,t)$ are thoroughly described by the equations (18), (20)–(22).

The stimulus current is:

$$I_s(t) = \begin{cases} I_s & 0 \le t \le t_1 \\ 0 & t > t_1 \end{cases} \tag{8}$$

where t_1 is the time duration of the pulse.

The boundary and initial conditions are given by the equations (9)–(14)

$$\lim_{x \to jL} V_z(x, t) = V_{zj}(x, t), j=1, 2, \dots, J \text{ for } z = a, m \tag{9}$$

where L is the length from nodal center to nodal center.

The model axon comprises 30 nodes and 29 internodes. Each internode is divided into two paranodal and five internodal segments. The lengths of node, paranode and nodal center to nodal center are 1.5, 200 and 1,400 μm, respectively. The length of each internodal segment is one fifth the overall length (998.5 μm) of the internode.

$$V_{mj}(t)=0 \text{ across the node and } V_j(t)=V_{aj}(t)+V_{mj}(t) \tag{10}$$

also

$$V_{z0}=V_{z2}; V_{z(J+1)}=0 \text{ for } z=a, m; \tag{11}$$

$$V(x, 0) = V_a(x, 0) = V_r \text{ across the node;} \tag{12}$$

as

$$V_a(x, 0) = V_{ra} \text{ across the internodal axolemma;}$$

and

$$V_m(x, 0) = V_r - V_{ra} \text{ across the myelin;}$$

so

$$V(x, 0) = V_r \text{ across the internode} \tag{13}$$

$$dz(x, 0)/dt = 0 \text{ for } z = V, V_a, V_m, m, h, n, s, q \tag{14}$$

In the paranodal space

$$R_{pa}(x) = R_{pn} \tag{15}$$

and in the periaxonal space

$$R_{pa}(x) = R_{pa} \tag{16}$$

In the nodal region:

$$C_a(x) = C_n \tag{17}$$

$$q_i(x, t) = I_{Na} + I_{Kf} + I_{Ks} \tag{18}$$

and in the internodal axolemma

$$C_a(x) = C_i \tag{19}$$

$$q_i(x, t) = I^*_{Na} + I^*_{Kf} + I^*_{Ks} + I^*_{IR} + I^*_{LK} + I^*_{pump} \tag{20}$$

where the *asterisk* denotes an internodal quantity and where

$$I_{Na} = P_{Na}m^3huz(Na); \quad I_{Kf} = P_{Kf}n^2z(K); \quad I_{Ks} = P_{Ks}sz(K);$$
$$I_{IR} = P_{IR}q(0.53z(Na) + 0.47z(K)); \quad I_{LK} = P_{LK}z(Na) \tag{21}$$

and

$$z(y) = (V_aF^2/RT)([y]_o-[y]_i \exp(V_aF/RT))/(1-\exp(V_aF/RT)) \tag{22}$$

for $y = $ Na, K ($T = 310.16K$; F, Faraday's constant; R, universal gas constant). The fraction activations are given by the differential equation:

$$dy/dt = a_y(1-y)-b_y y \text{ for } y = m, h, n, s, q \tag{23}$$

where

	A (ms^{-1})	B(mV)	C(mV)
a_m	1.872	−56.59	6.06
b_m	3.973	−61	9.41
a_h	0.550	−109.74	9.06
b_h	22.61	−26	12.5
a_n	0.129	−53	10
b_n	0.324	−78	10
s	0.00556	−60	22
q	0.00125	−110	−12
u		−80	−12

and

$$a_m, a_n = A(V_a-B)/(1-\exp((B-V_a)/C));$$
$$b_m, a_h, b_n = A(B-V_a)/(1-\exp((V_a-B)/C));$$
$$b_h = A/(1 + \exp((B-V_a)/C));$$
$$a_s, a_q = A \exp((V_a-B)/C);$$
$$b_s, b_q = A/(\exp((V_a-B)/C));$$
$$u = 0.7/(1+\exp((B-V_a(x,0))/C)) \tag{24}$$

In the internodally demyelinated space

$$R_{my}(x)=(R_{my}N(n(y\%)))/N(100\%) \text{ for } y\%–\text{myelin reduction value} \tag{25}$$

$C_{my}(x) = (C_{my} N(100\%))/N(n(y\%))$ for y%–myelin reduction value (26)

and $N(n(y\%)) = N(100\%)-N(y\%)$

where $N(n(y\%))$ is the number of myelin lamellae in the demyelinated zones.

In the paranodally demyelinated space

$$R_{pa}(x) = R_{pn}(x) \tag{27}$$

$$R_{pn}(x) = R_{pn}(100\%)-R_{pn}(y\%) \text{ for y\%–myelin reduction value.} \tag{28}$$

Other membrane parameter values for the normal motor nerve fibre are the same as described by Stephanova and Bostock (1995), i.e., C_n (nodal capacitance) 1 pF; C_i (internodal axolemmal capacitance) 350 pF; C_{my} (myelin capacitance) 2 pF; V_r (nodal resting potential) –86.7 mV; V_{ra} (internodal axolemmal resting potential) –86 mV; R_{ax} (axoplasmic resistance) 8 MΩ; R_{my} (myelin resistance) 250 MΩ. The channel types and their maximum permeabilities (cm³s⁻¹ x 10⁻⁹) are: node, Na (sodium) 9; K_f (fast potassium) 0.07; K_s (slow potassium) 0.26; internode, Na* (sodium) 80; K_f^* (fast potassium) 27; K_s^* (slow potassium) 2; IR* (inward rectifier) 0.008; L_k^* (leak) 0.0064; I_{pump}^* (net outward current generated by electrogenic Na⁺/K⁺ pump) 0.1 nA. The ion concentrations (mM) are: $[Na^+]_i$ 9; $[Na^+]_o$ 144.2; $[K^+]_i$ 155; $[K^+]_o$ 3. The presented maximum permeabilities of the channel types P_{Na} 9; P_{Kf} 0.07; 27*; P_{Ks} 0.26; P_{IR} 0.008* are the updated ones taken from the paper of Stephanova and Mileva (2000).

To clarify the relationship of the network topology (Fig. 1) to the morphology of the myelin, a schematic paranode-internode-paranode circuit in Fig. 3 ("real case") is presented to emphasize the curving of the longitudinal conducting myelin paths to meet the axon surface.

Electron micrographs indicate that each turn of the terminal cytoplasmic cord (TCC) corresponds to the termination of one myelin lamella (Berthold and Rydmark 1983). The turns of the TCC which belong to the 3–10 outer- and innermost myelin lamellae contain a diffuse material, however, 80–90% of the TCC contains a more electron-dense material. The presence of the TCC and its arrangement outside the MYSA axon explains the so-called "Spiny Bracelet of Nageotte". In Fig. 3 ("real case") this structural

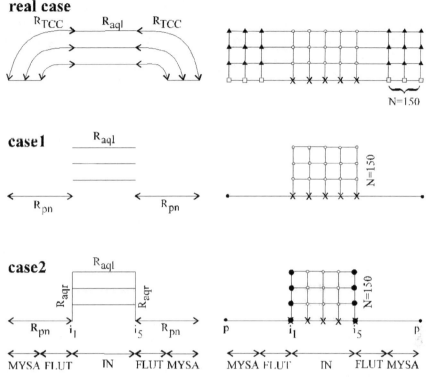

Fig. 3. Schematic paranode-internode-paranode circuits representing the *real case* and two limiting cases *case1*, *case2* of the morphological arrangement of the attachment of the ends of the longitudinally conducting myelin paths to the axon surface (*left-hand column*). The structural anisotropy of the myelin sheath for the same cases is shown in the *right-hand column*. Abbreviations: MYSA, paranodal segment (3–4 μm long) characterized by the myelin sheath attachment to the axolemma; IN, internode; FLUT, paranodal main segment characterized by a fluted axon; R_{pn}, paranodal seal resistance; R_{aql}, aqueous-layered longitudinal resistance; R_{TCC}, resistance of the terminal cytoplasmic cord; p, i_1 and i_5 indicate the positions of the paranodal segment and internodal first and fifth segments respectively. The successive symbol positions (for N=1,150) on the spiral radius vectors indicate, that there are distributed parameters as follows: o C_{myN}, R_{myN}, R_{aqlN}; • C_{myN}, R_{aqlN}, R_{myN}^* (* indicates that the R_{aqrN} chain is in parallel with the direct radial (spiral) R_{myN}); x R_{pa}^* (* indicates that the R_{aql1} chain is in parallel with R_{pa}, the periaxonal resistance); □ R_{pn}^* (* indicates that the R_{TCCN} chains are in parallel with the R_{pn}); ▲ R_{TCCN} (from Stephanova 2001).

organization is represented by constructing R_{pn} (in the MYSA section) as 150 segments in series and connecting each junction through a R_{aql} resistor to the end of a corresponding longitudinal conducting path. The latter is represented by the R_{TCC} resistor. Since such a circuit is entirely resistive, the calculations are simplified, and two limiting cases are modeled as "case 1" and "case 2". In the first one, the conducting paths within the myelin sheath end blindly at the paranodal ends, so that the longitudinal myelin currents are limited to those that have penetrated the myelin sheath, through the high resistance or small capacitance of the lipid layers. In this "case 1", the values of R_{TCC} are made infinite, so that the longitudinally conducting paths within the myelin end blindly at the paranodal periaxonal ends. This case is equivalent to a first case of the multi-layered model (Fig. 1). In the "case 2", the longitudinal current paths for the aqueous layers end in continuity with the paranodal periaxonal space. In this second case, the aqueous layers, converging at the paranodal ends, are presented as a chain of resistances (the entire R_{aqr} is defined as a radial myelin resistance). In the "case 2", the R_{TCC}'s values are set to a selected value of R_{aqr}, so that the longitudinally conducting paths within the myelin terminate simultaneously at the paranodal periaxonal ends. Such R_{aqrN} (for N=1,150) chain of resistors can provide a path for current flow separate from R_{pn}, if it is assumed that the core of the helical Spiny Bracelet presents very high resistance to current flow and if all of the teardrop-shaped "terminal cytoplasmic pockets" are joined side-to-side by high-resistance material (Fig. 1). The shown cases of structural anisotropy of the myelin sheath (Fig. 3, right-hand column) are based on the used helical sheet geometry of the sheath. Cross-sections at each point along the myelin sheath define spirals. The equation of the spiral in polar coordinates is $r = 2\pi.d/2 + \omega.\varphi_N$ where $\varphi_N \in (0, 2\pi.N)$ for N=1, 150 and $\omega = (D-d)/nl.N = 16$ nm for N = 150 is the lamella thickness (nl = 2 is the number of layers of the myelin lamellae). The successive points (i.e., r_N for N = 1, 150) on the radius vector are at a constant distance ω apart. The successive parameter values of R_{TCCN}, R_{aqrN}, R_{aqlN}, R_{myN} and C_{myN} (for N=1, 150), which define the structural anisotropy of the myelin sheath along its length, represent each turn of the spirals. The constant parameter values R_{TCC1}, R_{aqr1}, R_{aql1}, R_{my1}, C_{my1}, which define one lamella at a given point along the myelin sheath, represent one turn of the spirals. For the "real case", the conducting cytoplasmic cords (R_{TCCN}) provide paths for current flow in parallel with R_{pn} and connect the entire outer surface of the Schwann cell, i.e., the basement lamina with the axon surface. For "case 2", the conducting aqueous layer (R_{aqr}) in parallel with the distributed direct lipid layer (R_{my}) provides a current path separate from R_{pn} and connects the paranodal periaxonal space with the basement lamina.

The multi-layered model (Fig. 1) is equivalent to the double cable model (Fig. 2) when the aqueous layers within the myelin sheath are not taken into account; and in this case, the characteristic membrane parameter values for the normal human motor nerve fibre are: R_{aqr} (radial myelin resistance) ∞ MΩ; R_{aql} (longitudinal myelin resistance) ∞ MΩ; R_{pa} (periaxonal resistance) 300 MΩ; and R_{pn} (paranodal seal resistance) 125 MΩ. The same membrane parameter values are: R_{aqr} = 436.8 MΩ, R_{aql} = 21.3 MΩ, R_{pa} = 1250 MΩ and R_{pn} = 140 MΩ when the myelin aqueous layers are taken into account (Fig. 1). The membrane parameter values used in our model are adjusted to match both the recordings of threshold electrotonus (Bostock et al. 1991, 1994) and the recordings of action potentials (Dioszeghy and Stålberg 1992) in human motor nerves. The multi-layered model produces electrotonic potential indistinguishable from that of the double cable model with a periaxonal space 20 nm wide, and a resistivity of 70 Ωcm. The model axon diameter and the other geometric parameters presented here are the same as those used in most subsequent models of myelinated nerve fibres (Blight 1985, Awiszus 1990, Halter and Clark 1991).

The full system of differential equations is solved by the implicit numerical integration method (Euler version with subcycles) with a fixed time increment, usually 0.0001 or 0.005 ms for the calculation of the action or electrotonic potentials, respectively. Non-uniform spatial step sizes are used in accordance with the complex structure of the fibre. These step sizes are defined as lengths of the consecutive nodal, paranodal and internodal segments. They are not changed during the course of the computation. In the Euler version with subcycles, that we used, the time increment is automatically halved if the solution is unstable, and the given computation is repeated. This cycle can be repeated until the solution reaches numerical stability. Halving the internodal spatial steps and time steps has a negligible effect on the action potential parameters (i.e., amplitudes, durations and conduction velocities).

The multi-layered model, without or with taking into account the aqueous layers within the myelin sheath, is used for human motor nerve fibres in normal and pathological cases, such as in simulated cases of uniform axonal dysfunctions or in simulated cases of internodal, paranodal and simultaneously of paranodal internodal demyelinations, each of them systematic or focal. A given reduction of the myelin lamellae (defining internodal demyelination), or of the paranodal seal resistance (defining paranodal demyelination), or simultaneously both of them (defining paranodal internodal demyelination) is uniform along the fibre length for the systematically demyelinated subtypes. These permutations are termed internodal systematic demyelination (ISD), paranodal systematic demyelination (PSD) and paranodal internodal systematic demyelination (PISD), respectively. The reduction of the same myelin sheath parameters is used but restricted to only three (8th, 9th and 10th) consecutive internodes in the focally demyelinated cases. Such demyelinations are termed internodal focal demyelination (IFD), paranodal focal demyelination (PFD) and paranodal internodal focal demyelination (PIFD), respectively.

Our distributed parameter model can calculate the transaxonal, transmyelin, external membrane and ionic currents in each nodal, paranodal and internodal segment along the fibre length similar to the Halter and Clark model (1991). It is well known that a mixed nerve has a variety of fibre diameters and that the diameter of the underlying axon is an important determinant of intracellular action potential propagation. The geometric and membrane parameter values are in accordance with the mean axon diameter used in our model. This makes it possible to characterize, compare and analyze for systematic and focally demyelinated subtypes defining different demyelinating neuropathies.

Line-Source Model

Extracellular recordings can provide information about the bioelectric activity in a specific region of tissue. In that sense, the description of the extracellular potential field of a single fibre in an extensive conducting media is of great interest in the field of electrophysiology. A line-source model is the most frequently used model for calculating extracellular potentials (Clark and Plonsey 1968, 1970, Rosenfalck 1969, Plonsey 1977, Stegeman et al. 1979, Johannsen 1986, Ganapathy and Clark 1987, Stephanova et al. 1989, Trayanova et al. 1990, Strujik 1997, Meler et al. 1998). The set of equations describing the extracellular potential is given below and is taken from the paper of Stephanova et al. (1989).

From a rigorous solution of Laplace's equation, based on the theory of an active nerve fiber placed in an unbounded volume conductor, the extracellular potential P (r, z) can be calculated using the following set of equations:

$$P\,(r,\,z) = -\frac{d^2\sigma_i}{8\sigma_0}\,F^{-1}[V_f(k)\,k^2\,K_0(|k|r)] \tag{29}$$

where $V_f\,(k)$ is the Fourier transform of the transmembrane potential:

$$F\,[V(z)] = V_f(k) = \int_{-\infty}^{\infty} V(z)\,e^{\,jkz}\,dz, \tag{30}$$

while the inverse Fourier transform is defined as:

$$F^{-1}[V_f(k)] = V(z) = \frac{1}{2\pi}\int_{-\infty}^{\infty} V_f(k)\,e^{-jkz}\,dk \tag{31}$$

where k is the spatial frequency. In the equation (29): d is the fibre diameter; K_0 is the modified Bessel function of second kind, zero order; $V(z)$ is the transmembrane potential; σ_0 and σ_i are the specific conductivities of the extracellular and intracellular media, $\sigma_i/\sigma_0 = 1$; r and z are the radial and axial coordinates of the field point in the cylindrical coordinate system.

In the case of the line source model, the approximate solution of Laplace's equation gives the following expression for the extracellular potential of the active nerve fibre in an unbounded volume conductor:

$$P\,(z) = -\frac{d^2\sigma_i}{8\sigma_0}\int_{-\infty}^{\infty}\frac{\partial^2 V(z)}{\partial z^2}\,\frac{1}{R}\,dz \tag{32}$$

Using this approximate solution, the extracellular potential can be interpreted, as a convolution of the $\dfrac{\partial^2 V(z)}{\partial z^2}$ and the weighting function $1/R = 1/(r^2 + z^2)^{1/2}$. The model-generated spatial and temporal transmembrane (action) potentials, calculated by the already described above multi-layered model of the normal and abnormal human motor nerve axon, without or with myelin aqueous layers, are then used as input to the line source model that allows the calculation of the corresponding spatial and temporal extracellular potentials at various radial distances in the surrounding volume conductor.

Methods of Stimulation and Calculation of the Potentials (Action, Electrotonic and Extracellular)

To make accurate inferences about the physiological mechanisms involved in electric stimulation, one must know which elements are stimulated. Two cases of motor nerve axon stimulation are considered by us. The action potential stimulation is simulated by adding a short (0.1 ms) rectangular current pulse to the center of the first node. This case of point application of current intra-axonally at the node closely approximates the effects of point application of current extra-axonally at the node and realizes a point fibre polarization. The action potentials in the case of adaptation (i.e., in the case of intracellular current application delivered simultaneously to the center of each internodal segment) are simulated by adding a long-duration suprathreshold depolarizing pulse. This second case closely approximates the effects of external surface stimulation with a large electrode and realizes a periodic kind of uniform fibre polarization. The periodic kind of uniform fibre polarization was first simulated by Stephanova and Bostock (1996) in the human motor electrotonus model, since the *in vivo* conditions for recording threshold electrotonus in man, where current is applied to the peripheral nerve via its high-resistance sheath and a 1 cm diameter surface electrode (Bostock et al. 1991), is thought to correspond more closely to periodic kind of uniform polarization than point polarization of the fibre. The electrotonic stimulation is simulated by adding a 100 ms subthreshold current to the center of each internodal segment and the case of periodic kind of uniform fibre polarization is also realized. Electrotonic potentials are calculated for polarizing currents, which correspond to 0.4 times the threshold for a 1 ms depolarizing current stimulus. The term "accommodation" (electrotonic potential) was first applied to nerve by Nernst (1908), who found that, if a stimulating current is applied slowly enough, the nerve will "accommodate" to it and will not be excited. In this sense the electrotonic potential is assumed to be a passive transient. It is commonly accepted practice for theoreticians of electrotonus to ignore the membrane current density of a nonlinear function of the membrane potential. However, accommodation is also used in a somewhat different sense to describe the processes responsible for limiting the discharge induced by a sustained, suprathreshold current pulse (Bostock 1995). In this study, any reduction in excitability or increase in threshold that occurs prior to spike initiation, caused by a prolonged subthreshold pulse, is termed "accommodation", while "adaptation" is used for the reduction in impulse frequency, usually leading to cessation of repetitive activity, caused by a prolonged suprathreshold pulse. These two different aspects of accommodation are also used in our studies. According to Hill's (1936) model of accommodation, an axon should be excited as easily at the onset of a depolarizing stimulus (early adaptation) as at the offset of a long hyperpolarizing current stimulus (anode break excitation). Accommodation, adaptation and anode break excitation are of biological importance in determining whether and how often axons discharge when excited by slowly changing prolonged stimuli. There exists an extensive older literature on accommodation in nerve in which a close parallel was found between the accommodative changes in the excitability and the electrotonic potential. This is noted by Rosenblueth (1941) and explored in considerable detail by Lorente de Nó (1947). More recently, a non-invasive threshold tracking technique was developed (Bostock and Baker 1988, Bostock et al. 1998, Kiernan et al. 2000) to measure "threshold electrotonus", which revives the methods of studying nerve accommodation. According to Bostock (1995), the term "threshold electrotonus" is coined for the changes in threshold produced by long-lasting stimuli, as the accommodative changes in threshold largely parallel the underlying electrotonic potentials.

Extracellular potentials are calculated from action potentials in the case of adaptation.

Methods for Calculation of the Strength-Duration Time Constants, Rheobasic Currents and Recovery Cycles

In the case of adaptation strength-duration and charge-duration curves, strength-duration time constants and rheobasic currents are investigated. The threshold stimulus duration is increased in 0.025 ms steps from 0.025 ms to 1 ms, to obtain the strength-duration curves. These curves of the simulated normal and abnormal human motor nerve fibres are not natural exponential expressions, and the charge-duration curves are not linear. Due to this, a polynomial function of degree 2 (transfer standard parabola), which relates threshold charge (Q) to stimulus duration (t), provides an accurate fit of the data: $Q = a_2[t^2+(a_1/a_2).t +a_0/a_2]$, where a_0, a_1, a_2 are the polynomial coefficients. The strength-duration time constant (chronaxie) is defined as the absolute value of the smallest square root of the function (i.e., only one of both direct intercepts of the regression curve on the duration axis has a biophysical sense and only this direct intercept will be shown on the charge-duration figures). The rheobasic current is defined as the final decreased threshold value, after which the action potential generation cannot be obtained with an increase in stimulus duration.

When two equal-duration stimuli are used in pairs, the action potential in response to the second (testing) stimulus in the refractory period may be greater or less than that to the first (conditioning) stimulus and depends on the conditioning-test intervals. To obtain the time course of recovery of the axonal excitability following a single threshold stimulus (recovery cycle), test stimuli of 1 ms duration were delivered at conditioning-test intervals of 2–100 ms after a threshold conditioning stimulus of 1 ms duration.

Chapter III

Simulated Demyelinating Neuropathies and Neuronopathies

Simulation of CMT1A, CIDP, CIDP Subtypes, GBS, MMN and ALS

Recently, nerve excitability properties (such as threshold electrotonus, strength-duration time constants, rheobasic currents, recovery cycles) have been measured in healthy subjects and patients with demyelinating neuropathies (CMT1A, CIDP, GBS, MMN) and neuronopathies such as ALS (Bostock et al. 1995, Mogyoros et al. 1998, Cappelen-Smith et al. 2001, Kuwagara et al. 2002, 2003, Nodera et al. 2004, Sung et al. 2004), using the noninvasive technique of threshold tracking. Similarly, in the last ten years, we have simulated progressively greater degrees of mild systematic (ISDs, PSDs, PISDs −20,50,70,80%), mild focal (IFD, PFD, PIFD−70%), severe systematic (ISD − 93%, PSD − 90%, PISD − 82%) and severe focal (IFD, PFD, PIFD − 96%) demyelinations for human motor nerve fibres. We have also investigated and discussed their multiple excitability properties (such as action, electrotonic and extracellular potentials, strength-duration time constants, rheobasic currents, and recovery cycles) (Stephanova and Daskalova 2005a,b, 2008, Stephanova and Alexandrov 2006, Stephanova et al. 2005, 2006a,b, 2007a,b, 2011a,b, Krustev et al. 2010, Stephanova 2010). These studies confirm that the transition from conduction slowing (mild demyelinations) to conduction block (severe demyelinations) of action potential leads to amplification of the degree of excitability property changes, as the direction of these changes is maintained. The studies also confirm that the mild ISDs, PSDs and PISDs are specific indicators for CMT1A, CIDP and CIDP subtypes, respectively, whereas the severe IFD and PFD, PIFD are specific indicators for GBS and MMN, respectively.

Three progressively greater degrees of AI S motor neuron disease were observed by Bostock et al. (1995), comparing threshold electrotonus responses to 100 ms polarizing current stimuli, recorded from patients and normal controls. Following the above authors' axonal ion channel dysfunction suggestions, we have mathematically simulated, investigated and discussed these three ALS subtypes (Daskalova and Stephanova 2001, Stephanova and Daskalova 2002, Stephanova 2006, 2010).

Using our multi-layered model of human motor nerve fibre, without taking into account the myelin sheath aqueous layers, simulations of normal case, mild systematic and severe focal demyelinations each of them internodal, paranodal and paranodal internodal are given below. Demyelinations that were investigated are associated with a corresponding loss of myelin end bulbs and myelin lamellae away from the axolemma (Fig. 4). The characteristic parameter values defining the simulated cases are given in Table 1. A 70% reduction value is not sufficient to develop a conduction block of action potential in the systematically demyelinated cases. Such demyelinations are regarded as mild. A 96% reduction value is the first degree of the focally demyelinated subtypes for achieving the conduction block in a single internode. Such demyelinations are regarded as severe.

Table 1. Membrane parameter values characteristic for human motor nerve fibres in the normal and demyelinated cases, when demyelinations are mild (70%) and severe (96%), respectively. N (number of myelin lamellae); R_{my} (myelin resistance); C_{my} (myelin capacitance); R_{pn} (paranodal seal resistance).

		N	R_{my} [MΩ]	C_{my} [pF]	R_{pn} [MΩ]
Normal		150	250	1.5	125
ISD	70%	45	75	5	125
PSD	70%	150	250	1.5	37.5
PISD	70%	45	75	5	37.5
IFD	96%	6	10	37.5	125
PFD	96%	150	250	1.5	5
PIFD	96%	6	10	37.5	5

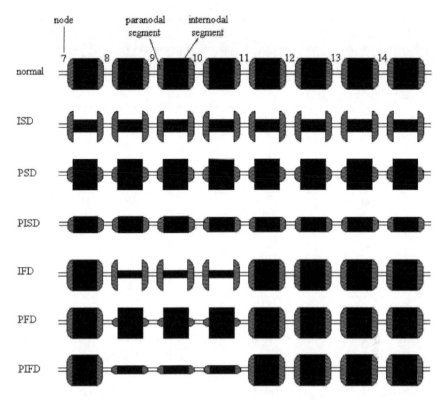

Fig. 4. Schematic diagram of human motor nerve fibres from the 7th to the 14th nodes in the normal, systematically and focally demyelinated cases. The reduction of the myelin lamellae (defining internodal systematic demyelination, ISD), or of the paranodal seal resistance (defining paranodal systematic demyelination, PSD), or simultaneously both of them (defining paranodal internodal systematic demyelination, PISD) is uniform along the fibre length. Reduction of the same myelin parameters is restricted to only three (8th, 9th and 10th) consecutive internodes for the internodal focal demyelination (IFD), paranodal focal demyelination (PFD) and paranodal internodal focal demyelination (PIFD), respectively (from Stephanova and Daskalova 2008).

We have also simulated cases of uniform nodal and internodal axonal dysfunctions termed ALS1, ALS2 and ALS3. Their characteristic parameter values are given in Table 2.

The mild systematic and severe focal demyelinations presented here are regarded as simulated hereditary (CMT1A), chronic (CIDP, CIDP subtypes) and acquired (GBS, MMN) demyelinating neuropathies, whereas the ALS subtypes presented here are regarded as simulated ALS motor neuron disease. The results

Table 2. Membrane parameter values characteristic for human fibres in the normal, ALS1, ALS2 and ALS3 cases. R_{ax} (axoplasmic resistance); R_{pn} (paranodal seal resistance); channel types and their maximum permeabilities: P_{Na} (sodium), P_{Kf} (fast potassium), P_{Ks} (slow potassium), P_{Lk} (leak). The *asterisk* denotes an internodal quantity. Values showing the differences between the normal and each ALS case are underlined.

cases	Normal	ALS1	ALS2	ALS3
$R_{ax}[M\Omega]$	8	<u>10</u>	10	10
$R_{pn}[M\Omega]$	125	<u>145</u>	145	145
$P_{Na}.10^{-9}[sm^3s^{-1}]$	9	<u>7</u>	7	7
$P_{Kf}.10^{-9}[sm^3s^{-1}]$	0.07	<u>0</u>	<u>0.07</u>	0.07
$P_{Ks}.10^{-9}[sm^3s^{-1}]$	0.26	<u>0.13</u>	<u>0</u>	0
$P_{Kf}*.10^{-9}[sm^3s^{-1}]$	27	<u>24.3</u>	<u>0</u>	0
$P_{Ks}*.10^{-9}[sm^3s^{-1}]$	2	<u>1.8</u>	<u>0.6</u>	<u>0</u>
$P_{Lk}*.10^{-11}[sm^3s^{-1}]$	0.64	<u>0.54</u>	<u>0.5</u>	0.5

presented here are consistent with the interpretation that genetic factors causing changes in the internodal segments of the myelin sheath could be responsible for the nerve excitability abnormalities exhibited in CMT1A. The results presented here are also consistent with the interpretation that immunological factors causing changes in the paranodal segments of the myelin sheath could be responsible for the clinical abnormalities exhibited in CIDP. Consequently, the same factors causing changes in demyelinations, which are heterogeneous (such as in some SIDP subtypes), could be responsible for the exhibited abnormalities in these diseases. Comparisons and analyses of the multiple excitability abnormalities obtained for the simulated neuropathies and ALS are given below.

Abnormalities in the Potentials

Action Potentials

Comparison of the action potentials is presented for the normal, ISD, PSD, PISD (Fig. 5, first column), IFD, PFD, PIFD (Fig. 5, second column), ALS1, ALS2 and ALS3 (Fig. 5, third column) cases of human motor nerve fibres. The potentials are of constant amplitude at the successive nodes for all systematically demyelinated and

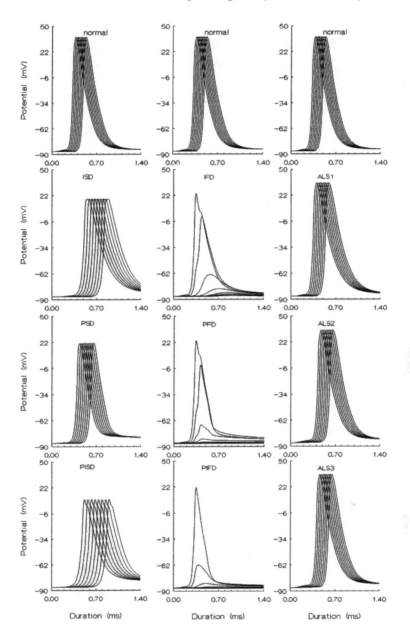

Fig. 5. Comparison between the action potentials in the normal (*first row*), mild ISD, PSD, PISD (*first column*), severe IFD, PFD, PIFD (*second column*), ALS1, ALS2 and ALS3 (*third column*) cases of human motor nerve fibres. The potentials in response to the applied 0.1 ms current stimuli are presented at each node from the 7th to the 14th.

ALS subtypes. Compared to the normal case, action potentials are with reduced amplitudes for the systematic demyelinations. They are near normal in all ALS cases, where the reduced amplitudes are negligible. Conduction velocities, calculated from the times of the potential maxima at the nodes are 58, 31, 45 and 25 m/s for the normal, ISD, PSD and PISD cases, respectively. They are 50, 51 and 51 m/s for the ALS1, ALS2 and ALS3 cases, respectively. Amplitudes are 18, 21 and 9 mV for the systematically demyelinated cases, respectively; and are 35, 35, 35 mV for the three ALS cases, respectively. The amplitude of the normal action potential is 38 mV. The progressively greater increase in focal loss of the myelin sheath blocks the invasion of potentials into the demyelinated zone (Fig. 5, second column). Thus, with the increase of the demyelination from IFD to PIFD, conduction failure occurs more rapidly.

The uniformly reduced amplitudes, prolonged durations and slowed conduction velocities of the action potentials shown here are an expected result for the investigated mildly systematically demyelinated subtypes. The conduction slowing simulated in the ISD, PSD, and PISD subtypes is a characteristic feature described in demyelinating neuropathies such as CMT1A, CIDP and its subtypes (Cappelen-Smith et al. 2001, Nodera et al. 2004, Sung et al. 2004).

A critical diagnostic feature of GBS and MMN is the demonstration of conduction block in multiple peripheral nerves on electrophysiological investigations (Kiernan et al. 2002, Kuwabara et al. 2002, Kaji 2003, Priori et al. 2005). Such a block was also demonstrated here in the IFD, PFD and PIFD cases. The simulated almost normal amplitudes and durations of action potentials and their near normal conduction velocities in the simulated ALS subtypes are in agreement with the slightly reduced velocities obtained for the almost normal action potentials recorded in patients with ALS (Sobue et al. 1983, Nakanishi et al. 1989, Wirguin et al. 1992, Arasaki and Tamaki 1998, Feinberg et al. 1999).

In routine diagnostic studies only latency or conduction velocity can be measured accurately. However, while such measurements may be very useful in defining pathology, they provide little insight into the underlying disease mechanisms, such as the mechanisms of conduction slowing, conduction block of the action potential or spontaneous impulse activity of the axons. Moreover, a number of morphological and functional changes such as branching, demyelination, remyelination, axonal tapering, axonal attenuation, axonal regrowth, cooling, axonal depolarization or hyperpolarization can affect the latency and conduction velocity of action potentials.

Mechanisms Defining the Action Potential Abnormalities in Simulated CMT1A, CIDP and CIDP Subtypes

The potentials in the simulated cases of ISD (CMT1A), PSD (CIDP) and PISD (CIDP subtypes) (Fig. 5, first column) are determined by their current kinetics (Fig. 6), which are compared to those in the normal case. The currents presented below are at node 10 only. To provide a better illustration: (1) the nodal ionic currents (I_{Na}, I_{Kf}, I_{Ks}, I_i) are presented in Fig. 6a; (2) the nodal transaxonal current (I_a, dotted line) and the nodal external membrane current (I_m) are presented in Fig. 6b and (3), all these currents are presented in Fig. 6c. The expected large inward current at the node of Ranvier resulting from the activation of a large number of nodal Na$^+$ channels can be seen in all investigated cases. In the normal, ISD, PSD and PISD cases, the nodal I_{Na} current (Fig. 6a,c) is activated rapidly by the membrane depolarization, and then it is inactivated. The contribution of nodal K$^+$ (fast and slow) channels to the membrane repolarization is less apparent in the ISD case, while in the PSD and PISD cases it is virtually absent. Consequently, action potentials at the nodal segments of simulated demyelinating neuropathies are determined mainly by the nodal sodium current (I_{Na}), as the contribution of nodal fast and slow potassium currents (I_{Kf} and I_{Ks}) to the total nodal ionic current (I_i) is negligible. Compared to the normal case, amplitudes of the nodal I_{Na} currents in the ISD and PISD cases are increased. At the node, the transmembrane potential (V_m) is composed of the transaxonal potential (V_a), and the "transmyelin" nodal gap potential. The latter is zero, as the resistive nodal gap is not taken into account in the model. However, the nodal transmembrane (external membrane) current (I_m, in Fig. 6b), for all investigated cases, is less than the current across the nodal axolemma (I_a). This is because the longitudinal current flows through the paranodal seal resistance (R_{pn}) to the periaxonal space.

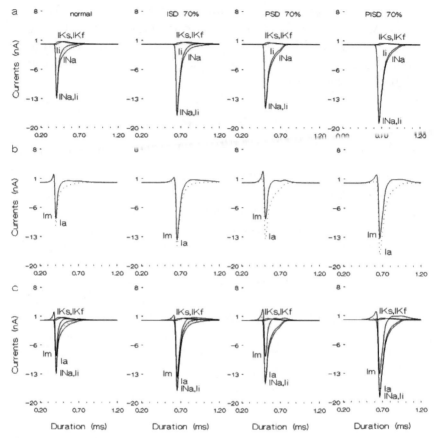

Fig. 6. Kinetics of the currents defining the action potential in the normal, ISD, PSD and PISD cases at node 10. Currents: **(a)** I_{Na} (sodium), I_{Kf}, I_{Ks} (fast, slow potassium), I_i (total ionic); **(b)** I_a (transaxonal, *dotted lines*), I_m (external membrane); **(c)** currents from **(a)** and **(b)** are given together.

Compared with the ISD case, the externally recorded nodal inward current (I_m, negative phase) in the PSD case is lower. The decrease of the paranodal seal resistance also results in an increase in the positive phase (reflecting the capacitive leakage current) of this external membrane current in the PSD and PISD cases.

Using our model, we can also distinguish between the transaxonal current across the internodal axolemma (I_a, in Fig. 7) and the transmembrane current (I_m, in Fig. 7) across the myelin sheath. In the paranodal and internodal segments, the external membrane current (I_m, in Fig. 7a,b) is equal to the transmyelin

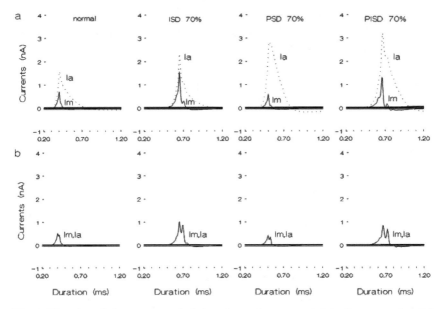

Fig. 7. Kinetics of the currents defining the action potential in the normal, mild ISD, PSD and PISD cases at paranode next to node 10 **(a)** and mid-internode between nodes 10 and 11 **(b)**. Currents: I_a (transaxonal, *dotted lines*), I_m (external membrane), I_{Na} (sodium), I_{Kf} (fast potassium), I_{Ks} (slow potassium), I_{IR} (inward rectifier), I_{Lk} (leak). In the paranodal and middle internodal segments, the external membrane current is equal to the transmyelin current. The I_{pump} current (0.1 nA) is not illustrated in the paranodal and middle internodal segments. The *straight lines* indicate the internodal ionic currents: I_{Na}, I_{Kf}, I_{Ks}, I_{IR}, I_{Lk}, respectively.

current. The transaxonal and transmyelin currents rapidly diminish in amplitude as the distance from the node increases, reaching equal minimal values in the center of the internode (I_a, I_m in Fig. 7b). An increase in amplitude of the transmyelin current across the paranodal axolemma can be seen in the ISD and PISD cases. However, considerably more outward current flows across the paranodal axolemma (I_a, in Fig. 7a) than is apparent from the current flowing across the myelin (I_m, in Fig. 7a) in the PSD and PISD cases. As in the normal case, for all demyelinated cases, the internodal ionic currents beneath the myelin sheath (I_{Na}, I_{Kf}, I_{Ks}, I_{IR}, I_{Lk}) are not changed significantly during the action potentials, either at the paranode and mid-internode. They appear as straight lines in Fig. 7a,b, respectively.

Compared with the normal case, the moment of the initial current flow through a given segment along the fibre length in all investigated systematically demyelinated cases, is prolonged. And as a result, the propagation of the action potential in the investigated demyelinated cases is delayed. The kinetics of the currents in the ISD case is consistent with the effect of the uniformly reduced myelin lamellae along the fibre length and shows the mechanisms of the conduction slowing of action potentials obtained in simulated CMT1A. The kinetics of the currents in the PSD case is consistent with the effect of the uniformly reduced paranodal seal resistance along the fibre length and shows the mechanisms of the conduction slowing of action potentials obtained in simulated CIDP. The kinetics of the currents in the PISD case is consistent with the effect of the reduced myelin lamellae, additionally increased by reduced paranodal seal resistance along the fibre length and shows the mechanisms of the conduction slowing of action potentials obtained in heterogeneous demyelinations, such as in some CIDP subtypes.

Mechanisms Defining the Action Potential Abnormalities in Simulated GBS and MMN

In the IFD (GBS), PFD and PIFD (MMN) cases, where conduction block of action potentials is realized (Fig. 6, second column) the currents discussed above are zero at the node 10, adjacent distal paranode, and mid-internode between nodes 10 and 11. Our previous data (Stephanova et al. 2007b) show that the current kinetics of the action potentials without conduction block in the same focally demyelinated cases compared to the current kinetics of the action potentials in the systematically demyelinated cases presented here is very similar for all investigated segments. However, in these focally demyelinated cases, the processes occur earlier. The kinetics of the currents, defining the mechanisms of conduction slowing of action propagation in simulated CMT1A, CIDP and its subtypes, was investigated for the first time in our previous study (Stephanova and Daskalova 2008). The results presented here confirm that the classical "transient" Na$^+$ current (100% in our simulations) contributes mainly to the action potential generation in the simulated demyelinating neuropathies. The internodal ion channels beneath the myelin sheath do not contribute to action potential generation at the paranodal and internodal segments along the fibre length. These ionic channels are insensitive to the short-duration current stimuli in simulated demyelinating neuropathies (CMT1A, CIDP, CIDP subtypes, GBS and MMN).

Mechanisms Defining the Action Potential Abnormalities in Simulated ALS

The action potential changes in the simulated cases of ALS1, ALS2 and ALS3 (Fig. 5, third column) are determined by the kinetics of the currents (Fig. 8), which are compared to those in the normal case. The investigated current kinetics of the action potentials in the three simulated ALS cases, is presented in the same manner as in Fig. 6, i.e., the shown currents are again at node 10 only; (1) the nodal ionic currents (I_{Na}, I_{Kf}, I_{Ks}, I_i) are presented in Fig. 8a; (2) the nodal transaxonal (I_a, dotted line) and external membrane (I_m) currents are presented in Fig. 8b; and (3), all these currents are presented in Fig. 8c. In the normal and abnormal cases, the comparison of action potential currents shows that they are quite similar (Fig. 8a). The activated nodal sodium channels (Na^+) dominate in the action potential generation realized by the short-duration (0.1 ms) current stimulus. The contribution of nodal slow potassium current (I_{Ks}) to the membrane repolarization is apparent in the ALS1 case, while in the ALS2 and ALS3 cases it is absent, as the slow K^+ channels are blocked there. Conversely, the contribution of nodal fast potassium current (I_{Kf}) to the membrane repolarization is apparent in the ALS2 and ALS3 cases, while in the ALS1 case it is absent, as the fast K^+ channels are blocked there. However, for all ALS cases, action potentials at the nodal segments are defined mainly by the nodal sodium current (I_{Na}), as the contribution of nodal fast and slow potassium currents (I_{Kf} and I_{Ks}) to the total nodal ionic current (I_i) is negligible. For the same cases, amplitudes of the nodal sodium currents are almost equal. Compared with the normal case, they are slightly reduced. The nodal external membrane current (I_m, in Fig. 8b) obtained for all ALS cases is less than the current across the nodal axolemma (I_a), because of the longitudinal current flow through the paranodal seal resistance (R_{pn}) to the periaxonal space.

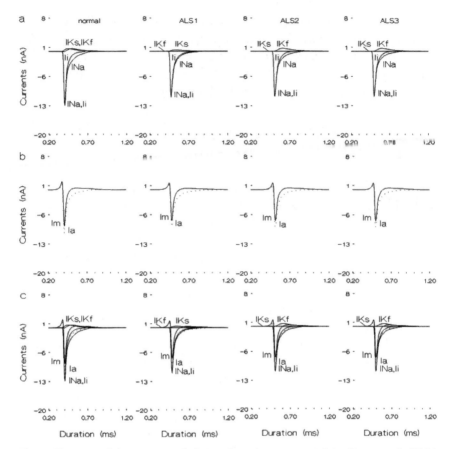

Fig. 8. Kinetics of the currents defining the action potential in the normal, ALS1, ALS2 and ALS3 cases at node 10. Currents: **(a)** I_{Na} (sodium), I_{Kf}, I_{Ks} (fast, slow potassium), I_i (total ionic); **(b)** I_a (transaxonal, *dotted lines*), I_m (external membrane); **(c)** currents from **(a)** and **(b)** are given together.

Compared with the normal case, amplitudes of the external nodal currents (I_m, negative phase) and those of the nodal transaxonal (I_a) currents are slightly reduced in all ALS cases. The increase of the axoplasmic (R_{ax}) and paranodal seal (R_{pa}) resistances (see Table 2 on page 36), in the ALS cases, results in a decrease of the longitudinal current flow through the paranodal axolemma to the periaxonal space as well as in a decrease of conduction velocities. As a result, amplitudes of the external nodal currents (I_m) and those of the nodal transaxonal currents (I_a) are reduced.

The transaxonal and transmyelin currents in the paranodal and mid-internodal segments (Fig. 9a) are almost unchanged in the ALS1, ALS2 and ALS3 cases. They are near normal. The amplitudes of the same currents diminish slightly as the distance from the node increases, reaching equal minimal values in the center of the internode (I_a, I_m in Fig. 9b). In the normal and abnormal ALS cases, the internodal ionic currents beneath the myelin sheath (I_{Na}, I_{Kf}, I_{Ks}, I_{IR}, I_{Lk}) do not change significantly during the action potential, as in the paranodal and mid-internodal segments. These currents are illustrated as straight lines in Fig. 9a,b. In the normal and abnormal ALS cases, the moments of the initial current flow through a given segment along the fibre length are almost the same, and as a result, the conduction velocities of action potentials in the abnormal cases are near normal. Action potential abnormalities depend on the

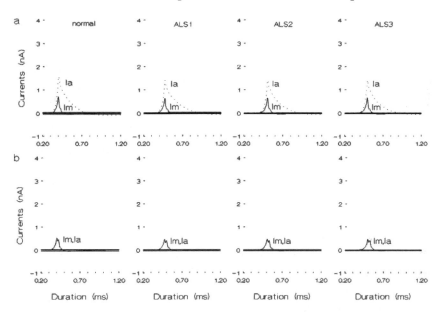

Fig. 9. Kinetics of the currents defining the action potential in the normal, ALS1, ALS2 and ALS3 cases at paranode next to node 10 **(a)** and mid-internode between nodes 10 and 11 **(b)**. Currents: I_a (transaxonal, *dotted lines*), I_m (external membrane), I_{Na} (sodium), I_{Kf} (fast potassium), I_{Ks} (slow potassium), I_{IR} (inward rectifier), I_{Lk} (leak). In the paranodal and middle internodal segments, the external membrane current is equal to the transmyelin current. The I_{pump} current (0.1 nA) is not illustrated in the paranodal and middle internodal segments. The *horizontal lines* are the internodal ionic currents: I_{Na}, I_{Kf}, I_{Ks}, I_{IR}, I_{Lk}, respectively.

configuration of the passive and active parameter values and on the ion channel dysfunction.

The equal values of the passive (R_{ax}, R_{pn}) and active (P_{Na}) parameters in all ALS cases (see Table 2 on page 36) and the small effect of blocked nodal K_f^+ channels in ALS1 and of blocked nodal K_s^+ channels in ALS2 and ALS3 (see Fig. 8a–c) on the action potential generation, explain the near equal conduction velocities (2% difference) between the propagating potentials through the ALS1 and ALS2, ALS3 axons, respectively. Conduction velocities are equal in the ALS2 and ALS3 cases, as the internodal ionic channels are insensitive to the short-duration current stimuli (see Fig. 9a,b). Consequently, the effect of the blocked fast (as in ALS1 case) or slow (as in ALS2, ALS3 cases) potassium nodal channels on the action potential repolarization will be insignificant (see Fig. 5 on page 37). The small effect of the blocked fast or slow potassium nodal channels on the action potential repolarization is a characteristic feature for normal human motor nerve axons (Schwarz et al. 1995, Stephanova & Mileva 2000). Regardless of the fact that the dysfunction of ion channels in the nodal and internodal axolemma progressively increases with each simulated ALS subtype, the action potentials cannot be regarded as specific indicators of the clinically recorded, progressively greater degrees of this disease. This result confirms that the investigation of action potentials as a routine diagnostic method may not be accurate.

Processes of action potential generation in the simulated ALS subtypes and simulated demyelinating neuropathies are quite different. Results show, that in simulated hereditary, chronic and acquire demyelinating neuropathies, abnormalities in action potentials and their corresponding current kinetics, based on the myelin sheath dysfunctions, are larger than those in the three simulated ALS subtypes, based on the progressively increased uniform axonal dysfunction.

Electrotonic Potentials

The electrotonic potentials are compared for the normal, ISD, PSD, PISD (Fig. 10, first column), IFD, PFD, PIFD (Fig. 10, second column), ALS1, ALS2 and ALS3 (Fig. 10, third column) cases of human motor nerve fibres. There are small differences in potentials

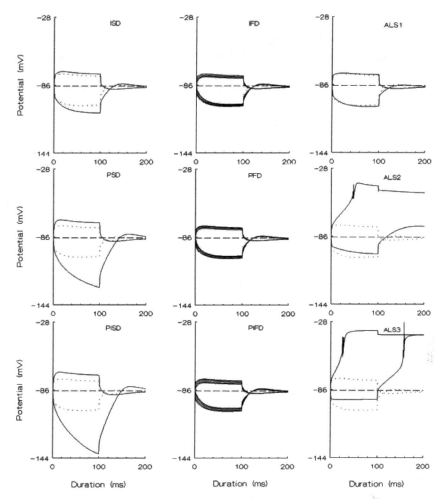

Fig. 10. Comparison between the electrotonic potentials in the normal (*dotted lines*), mild ISD, PSD, PISD (first column), severe IFD, PFD, PIFD (second column), ALS1, ALS2 and ALS3 (third column) cases of human motor nerve fibres. The potentials in response to 100 ms depolarizing and hyperpolarizing current stimuli (± 40% of the threshold) are presented at each node from the 7th to the 14th, respectively. However, each node behaves identically in the normal, systematically demyelinated and all ALS cases, and an overlap of the potentials at the nodes is obtained.

to the polarizing current stimuli between the normal and ISD cases. When the normal case is compared with the PSD and PISD cases, the obtained differences are the abnormally greater increase in the early and late parts of the hyperpolarizing responses (Fig. 10, first column).

For the focally demyelinated subtypes (Fig. 10, second column), electrotonic potentials are similar, with a small drop to a minimum amplitude in the 10th node and with a small rise to a maximum amplitude in the 14th node. There are almost no differences in electrotonic potentials during and after polarizing current stimuli for the normal and ALS1 cases (Fig. 10, third column). However, in the ALS2 case, spontaneous repetitive discharges are observed in the early part of the depolarizing response and attenuation of the hyperpolarizing response. In the ALS3 case, the axon is excited with less delay than in the ALS2 case, and it reproduces repetitive activities at the offset of the hyperpolarizing pulse as well as at the onset of the depolarizing pulse. The phenomenon anode break excitation [a.k.a. post-inhibitory rebound, (Hill 1936)] is observed in the ALS3 case after termination of the 100 ms hyperpolarizing current stimulus.

Simulated electrotonic potentials in the ISD case are rather similar to those described in demyelinating neuropathies, namely CMT1A (Nodera et al. 2004), whereas the abnormally increased hyperpolarizing responses in the PSD and PISD cases are similar to those described in CIDP and its subgroups (Sung et al. 2004). The electrotonic potential changes in the focally demyelinated subtypes are vastly smaller than those in the systematically demyelinated subtypes. There is a slight decrease in the electrotonic responses to subthreshold polarizing current stimuli in the IFD, PFD and PIFD subtypes. Compared with the normal threshold electrotonus, a smaller slow phase of threshold changes to polarizing current stimuli is recorded in AIDP patients (Kuwabara et al. 2002). This study on demyelinating forms of GBS without conduction block fails to reveal consistent abnormalities in the threshold electrotonus. Electrotonic potentials simulated here also fail to reveal consistent abnormalities in all focally demyelinated cases with conduction block. The near normal electrotonic potentials in the ALS1 case and the spontaneous discharges obtained during the early part of the depolarizing responses in the ALS2 and ALS3 cases are rather similar to those recorded in patients with ALS Type1, Type 2 and Type 3 (Bostock et al. 1995).

Electrotonic potentials allow the accommodative mechanisms to depolarizing and hyperpolarizing currents to be revealed. Observed changes in electrotonic potentials to the subthreshold depolarizing and hyperpolarizing current stimuli (Fig. 10), are determined by the current kinetics (Figs. 11–23) of the potentials. To provide a better comparison, the ionic currents in the nodal and internodal axolemma and transaxonal (I_a), transmembrane (I_m) currents are illustrated in different figures, however they will be discussed simultaneously. In each figure, the currents are presented for the normal case, for the given investigated cases and then they are compared.

Mechanisms Defining Abnormalities of the Polarizing Electrotonic Potentials in Simulated CMT1A, CIDP and CIDP Subtypes

During the depolarizing current stimuli, the temporal distributions of the ionic currents in the nodal and internodal axolemma (Fig. 11) and their corresponding transaxonal (I_a) and transmembrane (I_m) currents (Fig. 12), defining the depolarizing electrotonic potentials (Fig. 10, first column), are presented for the simulated cases of ISD (CMT1A), PSD (CIDP) and PISD (CIDP subtypes). The results show that the contribution of nodal slow potassium outward current (I_{Ks}) to both the total nodal ionic current (I_i, dotted line in Fig. 11a) and the generation of the nodal electrotonic potential in the normal and demyelinated cases, is obviously large. Other channels such as sodium (I_{Na}) and fast potassium (I_{Kf}) have negligible contribution to the potential generation at the node for the normal and ISD (CMT1A) cases.

Compared with the ISD case, the changes in the nodal ionic currents (I_{Ks}, I_{Kf} and I_{Na}) are bigger in the PSD case. They are biggest however, in the PISD case, as the effect of the paranodal demyelination is additionally increased by the effect of the internodal demyelination along the fibre length. Much of the externally recorded nodal current (I_m, continuous line in Fig. 12a) does not flow across the axolemma, since the I_a (dotted line) is much smaller, but passes longitudinally via the paranodal seal resistance to the periaxonal space. The decrease of the paranodal seal resistance causes an increase in the nodal external membrane current (I_m, PSD and PISD, in Fig. 12a). Conversely, considerably more currents flow across the paranodal and internodal axolemma (I_a, dotted

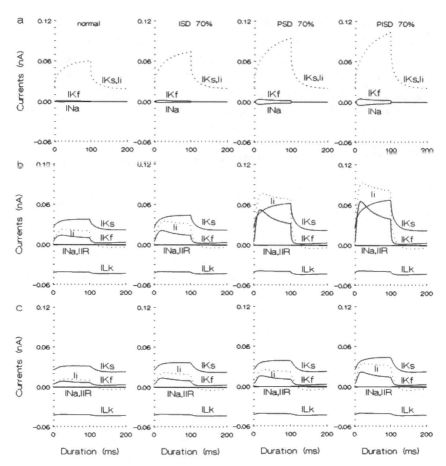

Fig. 11. During the subthreshold depolarizing current stimuli (+40% of the threshold), the kinetics of the ionic currents defining the depolarizing electrotonic potential in the normal, mild ISD, PSD and PISD cases is presented at node 10 **(a)**, paranode next to node 10 **(b)** and mid-internode between nodes 10 and 11 **(c)**. Currents: I_{Na} (sodium), I_{Kf} (fast potassium), I_{Ks} (slow potassium), I_i (total ionic, *dotted lines*) in the node; and I_{Na} (sodium), I_{Kf} (fast potassium), I_{Ks} (slow potassium), I_i (total ionic, *dotted lines*), I_{IR} (inward rectifier), I_{Lk} (leakage) in the internodal segments under the myelin sheath, respectively. The I_{pump} current (0.1 nA) is not illustrated in the paranodal and middle internodal segments.

line in Fig. 12b,c) than the current flowing across the myelin sheath (I_m, continuous line in Fig. 12b,c). Fast (I_{Kf}) and slow (I_{Ks}) potassium currents, resulting from the activation of ionic channels beneath the myelin sheath, dominate in the total ionic current (I_i, dotted line in Fig. 11b,c). Their amplitude changes in the paranodal segments are

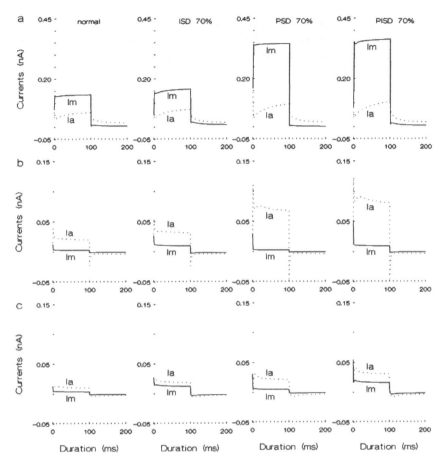

Fig. 12. During the subthreshold depolarizing current stimuli (+40% of the threshold), the kinetics of the transaxonal (I_a, *dotted lines*) and external membrane (I_m, *continuous lines*) currents defining the depolarizing electrotonic potentials in the normal, mild ISD, PSD and PISD cases is presented at node 10 **(a)**, paranode next to node 10 **(b)** and mid-internode between nodes 10 and 11 **(c)**. In the paranodal and middle internodal segments, the external membrane current is equal to the transmyelin current.

bigger in the PSD case and they are biggest in the PISD case (Fig. 11b). The paranodal demyelinations also cause an increase in the transaxonal current and a decrease in the transmyelin current at the paranodal and mid-internodal segments in the PSD and PISD cases (I_a, I_m, in Fig. 12b,c). I_{Kf}, I_{Ks} (Fig. 11c) and I_a, I_m (Fig. 12c) diminish, as the distance from the node increases, reaching minimal values in the center of the internode.

During the hyperpolarizing current stimuli, the ionic currents in the nodal and internodal axolemma (Fig. 13) and their corresponding transaxonal (I_a) and transmembrane (I_m) currents (Fig. 14), defining the hyperpolarizing electrotonic potentials (Fig. 10, first column), are presented for the simulated cases of ISD (CMT1A), PSD

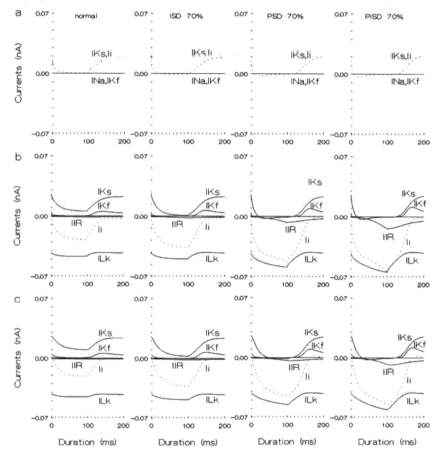

Fig. 13. During the subthreshold hyperpolarizing current stimuli (–40% of the threshold), the kinetics of the ionic currents defining the hyperpolarizing electrotonic potential in the normal, mild ISD, PSD and PISD cases is presented at node 10 **(a)**, paranode next to node 10 **(b)** and mid-internode between nodes 10 and 11 **(c)**. Currents: I_{Na} (sodium), I_{Kf} (fast potassium), I_{Ks} (slow potassium), I_i (total ionic, *dotted lines*) in the node; and I_{Na} (sodium), I_{Kf} (fast potassium), I_{Ks} (slow potassium), I_i (total ionic, *dotted lines*), I_{IR} (inward rectifier), I_{Lk} (leakage) in the internodal segments under the myelin sheath, respectively. The I_{pump} current (0.1 nA) is not illustrated in the paranodal and middle internodal segments. The *zero lines* in these internodal segments indicate internodal I_{Na} currents.

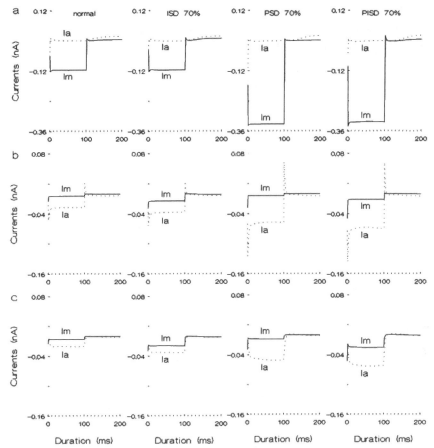

Fig. 14. During the subthreshold hyperpolarizing current stimuli (–40% of the threshold), the kinetics of the transaxonal (I_a, *dotted lines*) and external membrane (I_m, *continuous lines*) currents defining the hyperpolarizing electrotonic potential in the normal, mild ISD, PSD and PISD cases is presented at node 10 **(a)**, paranode next to node 10 **(b)** and mid-internode between nodes 10 and 11 **(c)**. In the paranodal and middle internodal segments, the external membrane current is equal to the transmyelin current.

(CIDP) and PISD (CIDP subtypes). The contribution of nodal ionic channels to the total ionic currents (I_i, dotted line) is negligible in the normal and abnormal cases (Fig. 13a). In the paranodal and mid-internodal segments, the internodal slow potassium (I_{Ks}) currents, in the normal and ISD cases, dominate in the total ionic currents (Fig. 13b,c). However, the activated inward rectifier (IR), leak (L_k) and internodal slow potassium channels, in the PSD and

PISD cases, dominate in the total ionic currents. Compared with the demyelinated cases during the depolarizing current stimuli (Fig. 12), the kinetics of the transaxonal (I_a) and transmyelin (I_m) currents, during the hyperpolarizing current stimuli presented here (Fig. 14), is rather similar for the different segments. It is noteworthy that during the hyperpolarizing current stimuli these currents flow in the opposite direction. The fast components of the external membrane (I_m) and transaxonal (I_a) currents are determined mainly by the passive cable responses, i.e., by the capacitances and resistances of the corresponding segments along the fibre length, and appear as vertical lines in Figs. 12 and 14.

The current kinetics of the electrotonic potentials described above in simulated CMT1A, CIDP and CIDP subtypes, shows that the slow components of potentials are dependent on the activation of channel types in the nodal or internodal axolemma, whereas the fast components of potentials are determined mainly by the passive cable responses, i.e., by the capacitance and resistance of the corresponding segments along the fibre length. Fast components of the transaxonal (I_a) and external membrane (I_m) currents reflect the moments of switching on and off of applied polarizing current stimuli. During the hyperpolarizing current stimulus, the contribution of activated internodal slow potassium (K^+), inward rectifier (IR) and leak (L_k) channels dominates in the total ionic currents in the paranodal and mid-internodal segments along the fibre length in simulated CIDP and its subtypes. However, during the depolarizing current stimuli, the contribution of activated potassium (K^+) channels in the internodal axolemma dominates in the total ionic currents. Furthermore, the depolarizing electrotonic potential at the node is determined mainly by the activation of the nodal slow potassium channels. For the paranodal and internodal segments, the fast and slow potassium currents, resulting from the activation of the ionic channels beneath the myelin sheath, dominate in the total ionic currents. For the same segments, the changes in the fast and slow potassium currents are biggest in the PISD case. Activated nodal sodium channels (Na^+), which dominate in the action potential generation realized by short-duration current stimuli, are not activated during the applied long-duration current stimuli. Similarly, the sodium channels in the internodal axolemma have a minor contribution to the generation of electrotonic potentials at the paranodal and mid-internodal segments. During the applied de- and hyperpolarizing current stimuli, the kinetics of the currents in the ISD case is consistent with the effect of the uniformly reduced myelin lamellae along the fibre length. And this shows the mechanisms of the accommodative processes defining the generation of electrotonic potentials obtained in simulated CMT1A. During the polarizing current stimuli, the kinetics of the currents in the PSD case is consistent with the effect of the uniformly reduced paranodal seal resistance along the fibre length. And this shows the mechanisms of the accommodative processes defining the generation of electrotonic potentials obtained in simulated CIDP. During the polarizing current stimuli, the kinetics of the currents in the PISD case is consistent with the effect of the uniformly reduced myelin lamellae, additionally increased by the effect of the uniformly reduced paranodal seal resistance along the fibre length. And this shows the mechanisms of the accommodative processes defining the generation of electrotonic potentials obtained in demyelinations, which are heterogeneous such as in some CIDP subtypes.

Mechanisms Defining Abnormalities of the Polarizing Electrotonic Potentials in Simulated GBS and MMN

During the depolarizing current stimuli, the temporal distributions of the ionic currents in the nodal and internodal axolemma (Fig. 15) and their corresponding transaxonal (I_a) and transmembrane (I_m) currents (Fig. 16), defining the depolarizing electrotonic potentials (Fig. 10, second column), are presented for the simulated cases of IFD (GBS), PFD and PIFD (MMN). Compared with the current kinetics in the systematically demyelinated cases (Figs. 11 and 12), the current kinetics in the focally demyelinated cases is quite similar. The contribution of nodal slow potassium current (I_{Ks}) to both the total nodal ionic current (I_i, dotted line in Fig. 15a) and the generation of the nodal electrotonic potential in the focally demyelinated cases, is again obviously large. Other channels such as sodium (I_{Na}) and fast potassium (I_{Kf}) again have negligible contribution to the potential generation at the node for all focally demyelinated cases. However, compared with the normal and

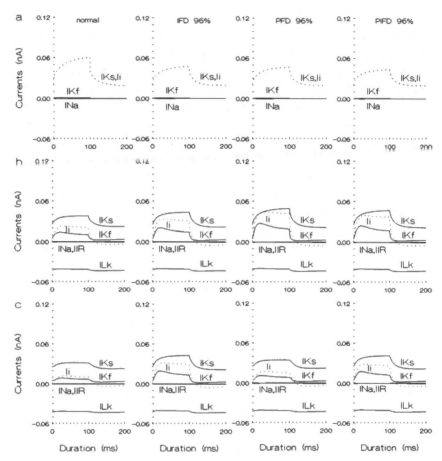

Fig. 15. During the subthreshold depolarizing current stimuli (+40% of the threshold), the kinetics of the ionic currents defining the depolarizing electrotonic potential in the normal, severe IFD, PFD and PIFD cases is presented at node 10 **(a)**, paranode next to node 10 **(b)** and mid-internode between nodes 10 and 11 **(c)**. Currents: I_{Na} (sodium), I_{Kf} (fast potassium), I_{Ks} (slow potassium), I_i (total ionic, *dotted lines*) in the node; and I_{Na} (sodium), I_{Kf} (fast potassium), I_{Ks} (slow potassium), I_i (total ionic, *dotted lines*), I_{IR} (inward rectifier), I_{Lk} (leakage) in the internodal segments under the myelin sheath, respectively. The I_{pump} current (0.1 nA) is not illustrated in the paranodal and middle internodal segments.

ISD, PSD and PISD cases, the almost equal nodal slow potassium currents in the IFD, PFD and PIFD cases are reduced.

Fast (I_{Kf}) and slow (I_{Ks}) potassium currents, resulting from the activation of ionic channels beneath the myelin sheath, dominate again in the total ionic current (I_i, dotted line in Fig. 15b,c). However, their changes are smaller in comparison to those in

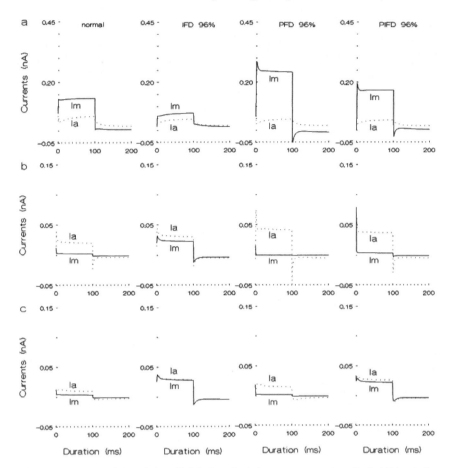

Fig. 16. During the subthreshold depolarizing current stimuli (+40% of the threshold), the kinetics of the transaxonal (I_a, *dotted lines*) and external membrane (I_m, *continuous lines*) currents defining the depolarizing electrotonic potentials in the normal, severe IFD, PFD and PIFD cases is presented at node 10 **(a)**, paranode next to node 10 **(b)** and mid-internode between nodes 10 and 11 **(c)**. In the paranodal and middle internodal segments, the external membrane current is equal to the transmyelin current.

the nodal segments for the PSD and PISD cases (Figs. 11b and 15b). There is a small difference between the current kinetics of the focally and systematically demyelinated cases at the middle internodal segments (Figs. 11c and 15c). The sodium (Na^+) and inward rectifier (IR) channels are not activated, whereas the leakage currents (I_{Lk}) are almost unchanged, during the long-duration depolarizing current stimuli. And this is the same for the systematically and for the focally demyelinated cases (Figs. 11b,c and 15b,c). Much of the externally recorded nodal current (I_m, in Fig. 16a) does not flow across the axolemma, since the I_a (dotted line) is much smaller, but passes longitudinally via the paranodal seal resistance to the periaxonal space. The decrease of the paranodal seal resistance causes an increase in the nodal external membrane current (I_m, PFD, in Fig. 16a), whereas the internodal demyelination causes a decrease in this current for the IFD and PIFD cases. Conversely, considerably more currents flow across the paranodal and internodal axolemma (I_a, dotted line in Fig. 16b,c) than is apparent from the current flowing across the myelin sheath (I_m, in Fig. 16b,c). However, compared with changes of the transaxonal and external membrane currents for the systematically demyelinated cases (I_a, I_m in Fig. 12a–c), changes in the same currents for the focally demyelinated cases (I_a, I_m in Fig. 16a–c) are considerably smaller.

During the hyperpolarizing current stimuli, the temporal distributions of the ionic currents in the nodal and internodal axolemma (Fig. 17) and their corresponding transaxonal (I_a) and transmembrane (I_m) currents (Fig. 18), defining the hyperpolarizing electrotonic potentials (Fig. 10, second column), are presented for the simulated cases of IFD (GBS), PFD and PIFD (MMN). Compared with the kinetics of the currents in the systematically demyelinated cases (Figs. 13 and 14), the current kinetics in the focally demyelinated cases is again quite similar (Figs. 17 and 18). For all focally demyelinated cases, the nodal slow potassium outward current (I_{Ks}) contributes to both the total nodal ionic current (I_i, dotted line in Fig. 17a) and the generation of the nodal electrotonic potential. Other currents such as nodal sodium (I_{Na}) and fast potassium (I_{Kf}) are again not activated during applied long-duration current stimuli. They appear as zero lines at the nodes. For all focally demyelinated cases, the internodal slow potassium (I_{Ks}) currents dominate in the total ionic currents (Fig. 17b,c) at the paranodal and mid-internodal segments. However, the contribution of activated inward rectifier (IR) and leak (L_k) channels to the total ionic currents is negligible, whereas the internodal sodium (I_{Na}) current is zero for all the investigated focally demyelinated cases.

The kinetics of the transaxonal (I_a) and external membrane (I_m) currents for the depolarizing current stimuli (Fig. 16a–c),

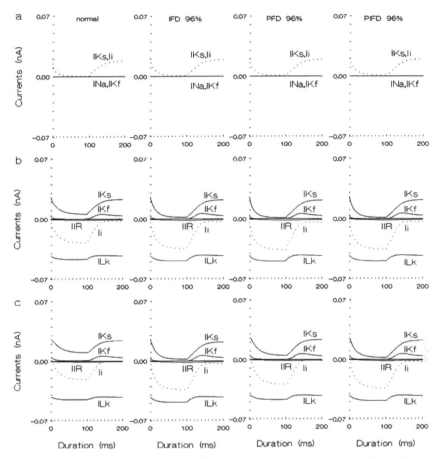

Fig. 17. During the subthreshold hyperpolarizing current stimuli (–40% of the threshold), the kinetics of the ionic currents defining the hyperpolarizing electrotonic potential in the normal, severe IFD, PFD and PIFD cases is presented at node 10 **(a)**, paranode next to node 10 **(b)** and mid-internode between nodes 10 and 11 **(c)**. Currents: I_{Na} (sodium), I_{Kf} (fast potassium), I_{Ks} (slow potassium), I_i (total ionic, *dotted lines*) in the node; and I_{Na} (sodium), I_{Kf} (fast potassium), I_{Ks} (slow potassium), I_i (total ionic, *dotted lines*), I_{IR} (inward rectifier), I_{Lk} (leakage) in the illustrated internodal segments under the myelin sheath, respectively. The I_{pump} current (0.1 nA) is not given in the paranodal and middle internodal segments. The *zero lines* in these internodal segments indicate I_{Na} currents.

compared with that for the hyperpolarizing current stimuli (Fig. 18a–c), is quite similar for the focally demyelinated cases. However, during the hyperpolarizing current stimuli, currents flow in the opposite direction. During polarizing current stimuli, the kinetics of the currents in the IFD case is consistent with the

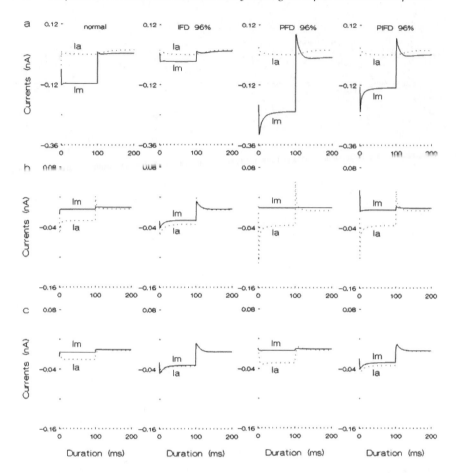

Fig. 18. During the subthreshold hyperpolarizing current stimuli (–40% of the threshold), the kinetics of the transaxonal (I_a, *dotted lines*) and external membrane (I_m, *continuous lines*) currents defining the hyperpolarizing electrotonic potential in the normal, severe IFD, PFD and PIFD cases is presented at node 10 **(a)**, paranode next to node 10 **(b)** and mid-internode between nodes 10 and 11 **(c)**. In the paranodal and middle internodal segments, the external membrane current is equal to the transmyelin current.

effect of the reduced myelin lamellae in the three (8th, 9th and 10th) consecutive internodes along the fibre length. And this shows the mechanisms of the accommodative processes defining the generation of electrotonic potentials obtained in simulated GBS. During polarizing current stimuli, the kinetics of the currents in the PFD and PIFD cases is consistent with the effect of both, the reduced paranodal seal resistance and the simultaneously reduced paranodal seal resistance and myelin lamellae, in the three (8th, 9th and 10th) consecutive internodes along the fibre length. And this shows the mechanisms of the accommodative processes defining the generation of electrotonic potentials obtained in simulated MMN. The results presented here confirm that the mechanisms underlying abnormalities in the polarizing electrotonic potentials for simulated hereditary (CMT1A), chronic (CIDP, CIDP subtypes) and acquired (GBS, MMN) demyelinating neuropathies are quite similar. However, in simulated GBS and MMN, current changes are smaller.

In the abnormally demyelinated cases, the comparison of the conduction slowing per block and accommodative processes shows that their mechanisms are different. Channel types beneath the myelin sheath show a high sensitivity to long-duration stimuli. However, the activated nodal sodium channels (Na+), which dominate in the action potential generation realized by short-duration current stimuli, are not activated during the long-duration current stimuli. Similarly, the internodal sodium channels have a minor contribution to the generation of electrotonic potentials. The described current kinetics of the electrotonic potentials show that the slow components of the potentials are dependent on the activation of channel types in the nodal or internodal axolemma, whereas the fast components of the potentials are determined mainly by the passive cable responses. The kinetics of the currents, defining the action or electrotonic potentials is similar in the systematically and focally demyelinated cases. However, in the focally demyelinated cases, the processes of conduction slowing or block are realized earlier. Compared with the systematically demyelinated cases, the current kinetics changes, defining the electrotonic potentials in the focally demyelinated cases, are relatively weak. The multiple investigations of the potential current kinetics, presented here in simulated demyelinating neuropathies, are based on our previous studies (Stephanova and Alexandrov 2006, Stephanova et al. 2007b, Stephanova and Daskalova 2008).

Mechanisms Defining Abnormalities of the Polarizing Electrotonic Potentials in Simulated ALS

For the depolarizing current stimuli, Figs. 19 and 20 show the trajectories of ionic currents in the nodal, internodal axolemma, and their corresponding transaxonal (I_a) and external membrane (I_m) currents in the simulated ALS1, ALS2 and ALS3 cases, defining the electrotonic potentials given in Fig. 10 (see page 47). Note that the y-scale of the last two columns is different in Figs. 19 and 20. There is an insignificant difference between normal and ALS1 current kinetics in the paranodal and mid-internodal segments. In the two kinetics, the fast (I_{Kf}) and slow (I_{Ks}) potassium currents, resulting from the activation of ionic channels beneath the myelin sheath, dominate in the total ionic current (I_i, in Fig. 19b). Considerably more currents flow across the paranodal and internodal axolemma (I_a, in Fig. 20b) which is apparent from the current flowing across the myelin sheath (I_m, in Fig. 20b). These currents slightly diminish, as the distance from the node increases, reaching minimal values in the mid-internode (Figs. 19c and 20c). The rectifier (IR) channels are not sensitive to the depolarizing stimuli; the internodal sodium (Na^+) channels are not activated because the fast and slow potassium channels are activated immediately at the onset of the depolarizing stimuli (zero lines, in Fig. 19b,c), whereas the leakage (I_{LK}) current is unchanged.

For the normal and ALS1 cases, the contribution of nodal slow potassium current (I_{Ks}) to both the total nodal ionic current (I_i, in Fig. 19a) and the generation of the nodal electrotonic potential is obviously large, whereas the other ionic channels such as sodium (Na^+) and fast potassium (K_f^+) are not activated. The latter are blocked in the ALS1 case. However, compared with the normal case, the slow potassium (I_{Ks}), total ionic current (I_i) and their corresponding transaxonal (I_a) and external membrane (I_m) currents are reduced in the ALS1 case (Figs. 19a and 20a).

The kinetics of the currents in the three ALS cases is quite different. In the ALS2 and ALS3 cases, the contribution of nodal sodium (I_{Na}) and fast potassium (I_{Kf}) currents to both the total nodal ionic current (I_i, continuous lines in Fig. 19a) and the generation of the nodal electrotonic potential is obviously large. The nodal

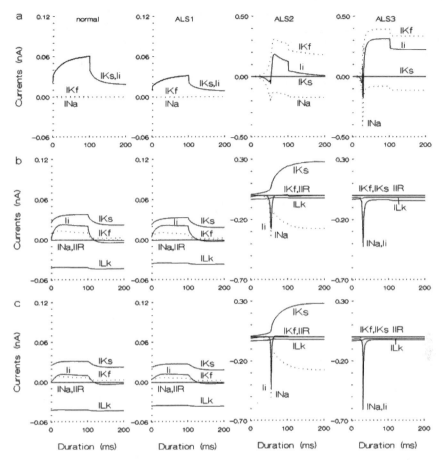

Fig. 19. During the subthreshold depolarizing current stimuli (+40% of the threshold), the kinetics of the ionic currents defining the depolarizing electrotonic potential in the normal, ALS1, ALS2 and ALS3 cases is presented at node 10 **(a)**, paranode next to node 10 **(b)** and mid-internode between nodes 10 and 11 **(c)**. Currents: I_{Na} (sodium, *dotted lines*), I_{Kf} (fast potassium, *dotted lines*), I_{Ks} (slow potassium) I_i (total ionic) in the node; and I_{Na} (sodium, *dotted lines*), I_{Kf} (fast potassium), I_{Ks} (slow potassium), I_i (total ionic), I_{IR} (inward rectifier), I_{Lk} (leakage) in the illustrated internodal segments under the myelin sheath, respectively. The I_{pump} current (0.1 nA) is not given in the paranodal and middle internodal segments. Note that the y-scales of the panel figures are different in the first two columns and last two columns.

slow potassium (K_s^+) channels are blocked in the ALS2 and ALS3 cases (see Table 2 on page 36). Moreover, inward currents at the nodes, resulting from the activation of a large number of nodal Na⁺ channels, can be seen in the ALS2 and ALS3 cases. The surge

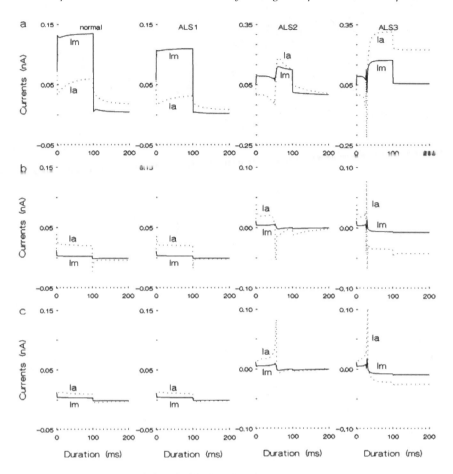

Fig. 20. During the subthreshold depolarizing current stimuli (+40% of the threshold), the kinetics of the transaxonal (I_a, *dotted lines*) and external membrane (I_m, *continuous lines*) currents defining the depolarizing electrotonic potentials in the normal, ALS1, ALS2 and ALS3 cases is presented at node 10 **(a)**, paranode next to node 10 **(b)** and mid-internode between nodes 10 and 11 **(c)**. In the paranodal and middle internodal segments, the external membrane current is equal to the transmyelin current. Note that the y-scales of the panel figures are different in the first two columns and last two columns.

of sodium current flow corresponds to the initiation of repetitive activity obtained in the early phase of the nodal total ionic current. This current increases from the ALS2 to the ALS3 case. The transients in the external membrane current (I_m) at the node are almost invisible in the ALS2 case and are apparent in the ALS3 case (Fig. 20a). The changes in the inward and outward phases of the transaxonal current (I_a) in the node are considerably larger in the ALS3 case than in the ALS2 case. Moreover, before and after initiation of the repetitive firing, considerably more currents flow across the nodal axolemma (I_a, dotted line in Fig. 20a) which is apparent from the current flowing across the myelin sheath (I_m, continuous line in Fig. 20a). In the ALS2 case, the total ionic currents (I_i), at the paranodal and mid-internodal segments depend on the activation of the internodal sodium (I_{Na}) and slow potassium (I_{Ks}) channels beneath the myelin sheath (Fig. 19b,c). Whereas, in the ALS3 case, the total ionic current (I_i) is equal to the sodium current (I_{Na}), as the slow potassium channels (K_s^+) beneath the myelin sheath are blocked. These internodal ionic currents increase, as the distance from the node increases, reaching maximal values in the center of the internode (Figs. 19c and 20c). The fast potassium channels (K_f^+) beneath the myelin sheath are blocked in the ALS2 and ALS3 cases (see again the Table 2). The inward rectifier (IR) channels are insensitive to the long-duration depolarizing current stimuli, whereas the leakage currents are slightly changed. However, compared to sodium current flow, these changes are vastly smaller (Fig. 19b,c). From the moment of initiation to the moment of cessation of the repetitive firing, considerably more currents flow across the internodal axolemma (I_a, in Fig. 20b,c), in the paranodal and mid-internodal segments, which is apparent from the current flowing across the myelin sheath (I_m, in Fig. 20b,c). However, the transaxonal (I_a) and transmyelin (I_m) currents flow in the opposite direction after the cessation of repetitive firing.

During the hyperpolarizing current stimuli, the ionic currents in the nodal and internodal axolemma (Fig. 21) and their corresponding transaxonal (I_a) and external membrane (I_m) currents (Fig. 22), defining the hyperpolarizing electrotonic potentials (Fig. 10, third column), are presented in the ALS1, ALS2 and ALS3 cases. Note that the y-scale of the last column is different in Figs. 21 and 22. The contribution of nodal slow potassium (K_s^+) channels to the total ionic currents (I_i) is negligible in the normal and ALS1 cases, whereas it is absent in the ALS2 cases, as the slow potassium channels (K_s^+) are blocked (Fig. 21a). The nodal sodium (Na^+) and fast potassium (K_f^+) channels are insensitive to the applied hyperpolarizing stimuli in the normal and abnormal ALS1, ALS2 cases. Moreover, the nodal fast potassium (K_f^+) channels are blocked in the ALS1 cases. For the normal and

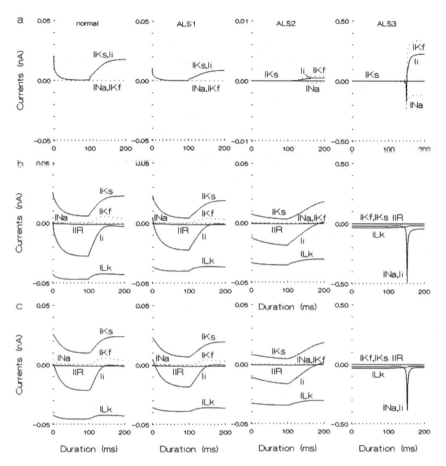

Fig. 21. During the subthreshold hyperpolarizing current stimuli (–40% of the threshold), the kinetics of the ionic currents defining the hyperpolarizing electrotonic potential in the normal, ALS1, ALS2 and ALS3 cases is presented at node 10 **(a)**, paranode next to node 10 **(b)** and mid-internode between nodes 10 and 11 **(c)**. Currents: I_{Na} (sodium, *dotted lines*), I_{Kf} (fast potassium, *dotted lines*), I_{Ks} (slow potassium) I_i (total ionic) in the node; and I_{Na} (sodium, *dotted lines*), I_{Kf} (fast potassium,), I_{Ks} (slow potassium), I_i (total ionic), I_{IR} (inward rectifier), I_{Lk} (leakage) in the illustrated internodal segments under the myelin sheath, respectively. The I_{pump} current (0.1 nA) is not given in the paranodal and middle internodal segments. Note that the y-scales of the panel figures are different in the first three columns and last column.

two ALS1 and ALS2 abnormal cases, the internodal slow potassium (I_{Ks}) and leakage (I_{LK}) currents, at the paranodal and mid-internodal segments, dominate in the total ionic currents (Fig. 21b,c). The total

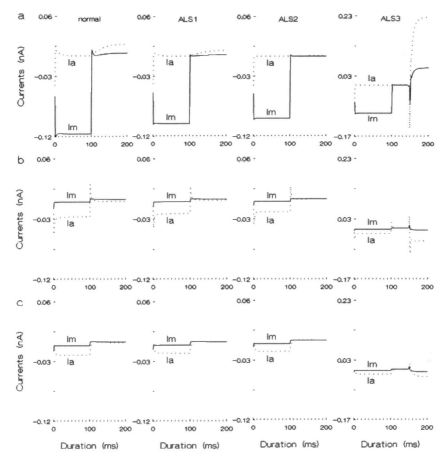

Fig. 22. During the subthreshold hyperpolarizing current stimuli (–40% of the threshold), the kinetics of the transaxonal (I_a, *dotted lines*) and external membrane (I_m, *continuous lines*) currents defining the hyperpolarizing electrotonic potentials in the normal, ALS1, ALS2 and ALS3 cases is presented at node 10 **(a)**, paranode next to node 10 **(b)** and mid-internode between nodes 10 and 11 **(c)**. In the paranodal and middle internodal segments, the external membrane current is equal to the transmyelin current. Note that the y-scales of the panel figures are different in the first three columns and last column.

ionic current (I_i) diminishes in the ALS2 case, as the slow potassium (I_{Ks}) and leakage (I_{LK}) currents are decreased.

The internodal sodium (I_{Na}) current changes are negligible to the applied hyperpolarizing stimuli, whereas the fast potassium (K_f^+) channels are blocked. Compared with the normal case, the kinetics of the transaxonal (I_a) and transmyelin (I_m) currents are very similar in the ALS1 and ALS2 cases during the long-duration depolarizing (Fig. 20) and hyperpolarizing (Fig. 22) stimuli. However, during the hyperpolarizing current stimuli, these currents flow in the opposite direction.

In the most abnormal ALS3 case, the current kinetics shows the initiation and generation of spontaneous action potential after the termination of the applied long-duration hyperpolarizing current stimuli (Fig. 21a–c). The action potential generation at the node depends on the activation of sodium (Na^+) and fast potassium (K_f^+) channels, as the slow potassium (K_s^+) channels are blocked there (Fig. 21a). The action potential generation, at the paranodal and internodal segments, depends mainly on the activation of sodium (Na^+) channels beneath the myelin, as the potassium channels (fast and slow) beneath the myelin sheath are blocked in the ALS3 case (see Table 2 on page 36). The inward rectifier (IR) channels are not activated, whereas the leakage current is slightly changed (Fig. 21b,c). However, compared to the sodium current flow, this current change is vastly smaller (Fig. 21b,c). The kinetics of the transaxonal (I_a) and external membrane (I_m) currents, during the applied hyperpolarizing current stimulus is the same as that in the ALS1 and ALS2 cases (Fig. 22a–c). Following the termination of the applied stimulus, the current kinetics in the ALS3 case shows spontaneous action potential generation in all segments along the fibre length. In this case, the contribution of "transient" Na^+ current to the generation of the repetitive firing in the polarizing electrotonic responses can be clearly seen on the expanded scales in Fig. 23.

Based on experimental data, as well as the theory of singularly perturbed systems of differential equations and the bifurcation theory of dynamic systems, it is well known (Rinzel 1987, Bertam et al. 1995, Wang and Rinzel 1995, Guckenheimer 1997) that electrophysiological systems can reproduce different types of bursting oscillations. Moreover, it was shown that there are many different ways in which single or combination blocked potassium channels can be altered to achieve different types of bursting oscillations in the potentials of simulated activity of human motor nerve fibres (Stephanova and Mileva 2000). The results presented here confirm that the configuration used for the characteristic passive (RC) and active (ionic channels) membrane parameter

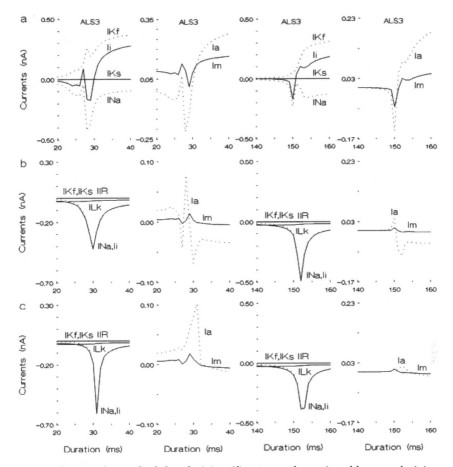

Fig. 23. During the applied depolarizing (first two columns) and hyperpolarizing (second two columns) current stimuli (±40% of the threshold), the kinetics of the ionic currents (first and third columns) and their corresponding transaxonal and external membrane currents (second and fourth columns) are presented in the ALS3 case on expanded scales. The ionic currents in the nodal and internodal axolemma as well as the y-scales of the panel figures are the same as in Figs. 19, 20, 21 and 22, respectively.

values (see Table 2 on page 36) can correctly reproduce both the burst activities in the electrotonic responses to the depolarizing and hyperpolarizing current stimuli obtained in the ALS2 and ALS3 cases and the near normal electrotonic responses to the polarizing current stimuli in the ALS1 case.

Results show that excitability property changes obtained in simulations replicate those recorded *in vivo* in patients with ALS Type1, Type2 and Type3. Impaired K^+ channels in ALS patients were proposed by Bostock et al. (1995), but this has not yet been established or confirmed by later studies. Consequently, the current kinetics of the polarizing electrotonic potentials for the simulated ALS2 and ALS3 subtypes confirm that the repetitive discharges, due to the progressively increased nodal and internodal ion channel dysfunction, are consistent with the activation of the nodal and internodal sodium channels as well as with the loss of functional potassium channels involving both the fast and slow potassium channel types.

During the applied polarizing current stimuli, the kinetics of the currents in the ALS1 case is consistent with the effect of the uniformly changed passive (R_{ax}, R_{pn}) and active (ion channel) parameters along the fibre length. And this shows the mechanisms of the accommodative processes defining the generation of the near normal electrotonic potentials obtained in the simulated ALS Type1. During the applied polarizing stimuli, the kinetics of the currents in the ALS2 case is consistent mainly with the effect of the blocked nodal slow (K_s^+) and blocked internodal fast (K_f^+) potassium channels along the fibre length. And this shows the mechanisms of the generation of the repetitive discharges in the early parts of the depolarizing electrotonic responses obtained in the simulated ALS Type2. After termination of the applied hyperpolarizing stimuli, the kinetics of the currents in the ALS3 case is consistent with the effect of the simultaneous blockage of the internodal fast and slow (K_f^+, K_s^+) potassium channels in combination with the blocked nodal slow (K_s^+) potassium channels, along the fibre length. And this shows the mechanisms of generation of the repetitive discharges (action potentials) in the later parts of the hyperpolarizing electrotonic responses obtained in the simulated ALS Type3. In the ALS2 and ALS3 cases, the comparison of the conduction slowing and accommodative processes shows that their mechanisms are quite similar. The nodal sodium channels (Na^+), which dominate in the action potential generation realized by short-duration current stimuli, are also activated during the applied long-duration subthreshold polarizing current stimuli. Similarly, sodium channels beneath the myelin sheath have a major contribution to the generation of the polarizing electrotonic potentials in the internodal segments. The current kinetics of the electrotonic potentials described in the three simulated ALS types shows that the slow components of potentials are dependent on the activation of the channel types in the nodal or internodal axolemma, whereas the fast components of potentials are determined mainly by the passive cable responses. The multiple investigations of the current kinetics defining the action and electrotonic potentials in the simulated ALS cases are based on our previous studies (Stephanova et al. 2012a,b). The electrotonic potentials in the simulated ALS are quite different from those in the simulated demyelinating neuropathies because of the different fibre electrogenesis.

Homogeneity or Heterogeneity of Membrane Polarization in Simulated Demyelinating Neuropathies without or with Conduction Block

Conduction block is an important functional consequence of demyelination whereby nervous transmission is abolished. Conduction block is defined as the failure of a nerve impulse to propagate through a structurally intact axon. Moreover, a number of morphological and functional changes such as demyelination, axonal channel blockage, branching, cooling, axonal de- or hyperpolarization can provoke conduction block. Among these mechanisms of conduction block, the most important is demyelinated conduction block, which accounts for much of the clinical deficits of demyelinating neuropathies. As already discussed, a demonstration of conduction block is essential for diagnosing MMN (Kaji 2003, Priori et al. 2005), GBS (Feasby et al. 1986, Chowdhury and Arora 2001), CIDP, etc. CIDP can occur with other systemic diseases (Barohn et al. 1989, Kuwabara et al. 1993, Katz et al. 2000).

A recent study (Kiernan et al. 2002) assessing accommodative processes consistently shows hyperpolarization of the axonal membrane outside the region of the conduction block in patients with MMN. No information is available on the membrane polarization inside the region of the conduction block. However, several indirect observations in MMN (Kaji 2003, Priori et al. 2005) suggest the idea that the abnormal membrane depolarization at the site of the conduction block of action potential in turn leads to a compensatory membrane hyperpolarization outside the block.

To complete the knowledge of the accommodative processes and to improve the knowledge of the membrane polarization based on these processes in the above mentioned demyelinating neuropathies, the spatial distributions of the electrotonic potentials of demyelinated human motor nerve fibres without or with conduction block of the action potential are investigated. Fibres in simulated cases of internodal, paranodal and simultaneously of paranodal internodal demyelinations, each of them systematic or focal, are studied. The study is performed for 70, 93, 89, 82 and 96% myelin reduction values. The first value is not sufficient to develop conduction block of the action potential in systematically demyelinated cases (ISD, PSD, PISD) as well as in focally demyelinated cases (IFD, PFD, PIFD) and these demyelinations are mild. The remaining three 93, 89, 82% values lead to conduction block in the ISD, PSD and PISD cases, respectively, whereas the last 96% value leads to conduction block in all focally demyelinated cases (IFD, PFD, PIFD), and the demyelinations are severe. A comparison of the temporal distributions of the polarizing electrotonic potentials is presented for the normal, ISD, PSD, PISD (Fig. 24a,b), IFD, PFD and PIFD (Fig. 24c,d) cases of human motor nerve fibres with mild (a, c) and severe (b, d) demyelinations.

The results show that the temporal electrotonic responses for each systematically demyelinated case without conduction block abnormally increase in the case of conduction block. In the focally demyelinated cases without or with conduction block, the small decrease of temporal electrotonic responses in the demyelinated zone in turn leads to a compensatory increase of these responses outside the demyelinated zone.

According to Kiernan et al. (2002), early and late depolarizing (TEd) and hyperpolarizing (TEh) threshold electrotonus parameters [early TEd (10–20 ms) (%); late TEd (90–100 ms) (%); early TEh (10–20 ms) (%) and late TEh (90–100 ms) (%)] are taken as the most appropriate parameters to reflect demyelination at the simulated site (i.e., outside the region of conduction block). These parameters measure the so-called early and late "fanning out" or "fanning in" of the threshold electrotonus responses and

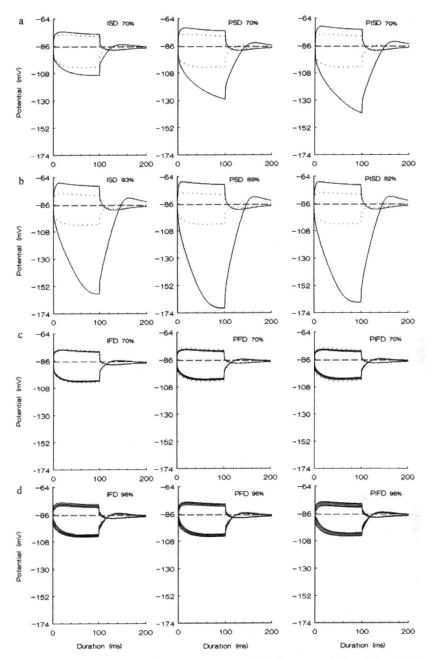

Fig. 24. Comparison of the electrotonic potentials in the normal (*dotted lines*), ISD, PSD, PISD **(a, b)**, IFD, PFD and PIFD **(c, d)** cases is presented with mild **(a, c)** and severe **(b, d)** demyelinations.

are dependent on the applied polarizing current stimuli and demyelinated factors. Analogically, early and late depolarizing (Ed) and hyperpolarizing (Eh) electrotonic parameters [early Ed (10–20 ms) (mV); late Ed (90–100 ms) (mV); early Eh (10–20 ms) (mV) and late Eh (90–100 ms) (mV) after the start of current stimuli] are also used for the electrotonic potentials presented here. The values of these parameters at each node (from the 1st to the 30th) give the spatial distribution of the electrotonic potential along the fibre length.

The spatial distributions of the polarized electrotonic potentials are plotted for the ISD, PSD and PISD subtypes without (Figs. 25a and 26a) and with (Figs. 25b and 26b) conduction block at the early (mean 15 ms) and late (mean 95 ms) moments. They are compared with those of the normal case. The results show that the abnormal increases ("fanning out") of the electrotonic responses, without and with conduction block, remain unchanged along the fibre length at the early (Fig. 25a,b) and late (Fig. 26a,b) moments, and these are shown as straight lines. The electrotonic potential for the normal case (dotted lines) is also unchanged along the fibre length.

The same investigations are repeated for the focally demyelinated subtypes. The spatial distributions of the potentials are presented for the normal, IFD, PFD and PIFD cases without and with conduction block at the early (Fig. 25c,d) and late (Fig. 26c,d) moments. For the focally demyelinated cases without conduction block, the decrease ("fanning in") of the polarizing electrotonic responses in the demyelinated location and its vicinities in turn leads to a compensatory recovery of these responses to the normal value outside the demyelinated zone (Figs. 25c and 26c). For the same focally demyelinated cases with conduction block, the decrease ("fanning in") of the polarizing electrotonic responses in the demyelinated zone in turn leads to a compensatory increase ("fanning out") of these responses outside the demyelinated zone (Figs. 25d and 26d).

The results show that the transition from conduction slowing to conduction block leads to amplification of the degree of electrotonic changes in systematically demyelinated cases (ISD, PSD, PISD), as the direction of these changes is maintained.

For the focally demyelinated cases (IFD, PFD, PIFD) with conduction block, the "fanning in" of the polarizing electrotonic

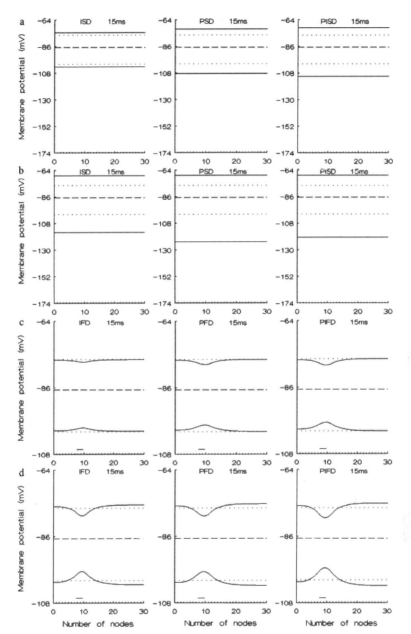

Fig. 25. Comparison between the early (15 ms) spatial potential distributions for the normal (*dotted lines*), ISD, PSD, PISD **(a, b)**, IFD, PFD, and PIFD **(c, d)** cases is presented when the demyelinations are mild **(a, c)** and severe **(b, d)**, respectively. The *dashed lines* indicate the resting potential, whereas the *short line* (**c, d,** at the bottom) indicates the location of the demyelination.

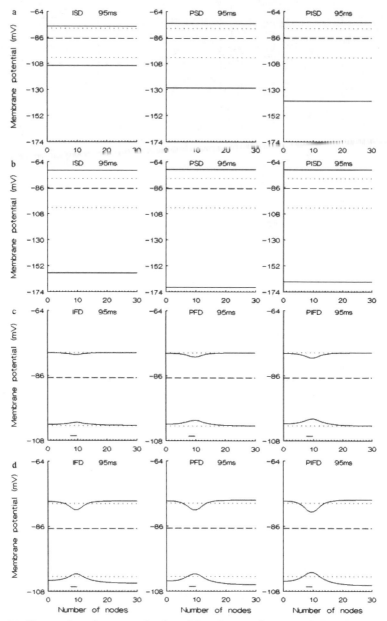

Fig. 26. Comparison between the late (95 ms) spatial potential distributions for the normal (*dotted lines*), ISD, PSD, PISD (**a, b**), IFD, PFD, and PIFD (**c, d**) cases is presented when the demyelinations are mild (**a, c**) and severe (**b, d**), respectively. The *dashed lines* indicate the resting potential, whereas the *short line* (**c, d,** at the bottom) indicates the location of the demyelination.

responses in the demyelinated zone in turn leads to a compensatory "fanning out" of these responses outside the demyelinated zone. These results confirm that the membrane is depolarized at the site of the conduction block and that the membrane is hyperpolarized outside the region of the conduction block in the simulated focally demyelinated subtypes.

Consequently, based on the spatial distributions of the electrotonic responses to the polarizing current stimuli, mathematical evidence is given for the heterogeneity of the membrane polarization in the case of conduction block in simulated acquired demyelinating neuropathies such as GBS and MMN. Moreover, the spatial distributions of the electrotonic responses to the polarizing current stimuli also confirm homogeneity of the membrane polarization in the case of conduction block in simulated hereditary and chronic demyelinating neuropathies such as some types of CMT1A and CIDP. The results presented here are based on our previous investigations (Stephanova and Alexandrov 2006, Stephanova and Daskalova 2009).

Abnormalities in the Extracellular Potentials and their Mechanisms

Figure 27 illustrates the temporal distributions of the action potentials (V) and their corresponding extracellular potentials (P) in the case of adaptation for the normal, mildly systematically (first panel), severely focally (second panel) demyelinated and the three ALS (third panel) cases. The extracellular potentials have the usual three-phase shape in the normal case. The potentials show increased polyphasia in the PSD and PISD subtypes (Fig. 27, first panel). The peak-to-peak amplitude of the potentials decreases in the demyelinated zone of the severely focally demyelinated cases (Fig. 27, second panel), and this can be seen more clearly in Fig. 28. In this figure, the extracellular potential time courses are taken from Fig. 27. The potentials are plotted for the same eight consecutive nodes from the 7th to the 14th as shown in the first row, except in the second, third, fourth and fifth rows where the potentials are shown at the 7th, 8th, 9th and 10th nodes, respectively. These results show that with an increase of the demyelinated subtypes from the IFD to the PIFD, the peak-to-peak amplitude of the potentials abnormally

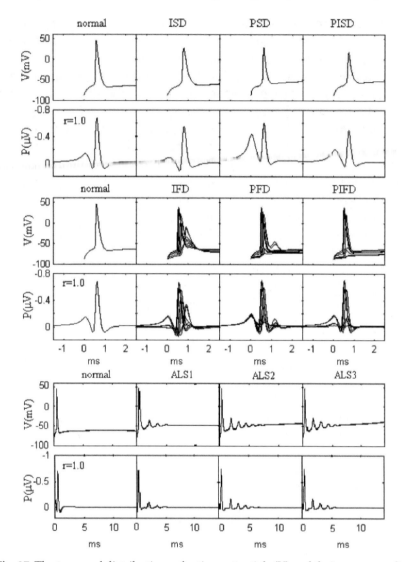

Fig. 27. The temporal distributions of action potentials (V) and their corresponding extracellular potentials (P) in the normal, mild ISD, PSD, PISD (*first panel*), severe IFD, PFD, PIFD (*second panel*), ALS1, ALS2 and ALS3 (*third panel*) cases are presented for human motor nerve fibers. The action potentials are simulated by adding long-lasting depolarizing stimuli, which correspond to 1.1 times the threshold for a 1 ms current stimulus. The potentials are at each node from the 7th to 14th. However, each node behaves identically in the normal, systematically demyelinated and in all ALS cases, and an overlap of the potentials at the nodes is obtained. The extracellular potentials are given for a radial distance r = 1.0 mm.

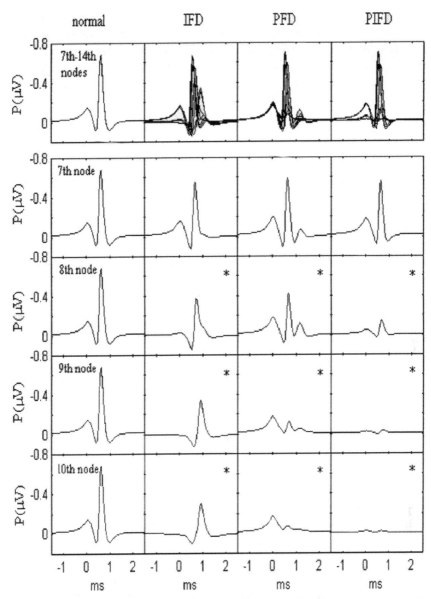

Fig. 28. The temporal distributions of extracellular potentials (P) in the normal, IFD, PFD and PIFD cases are repeated from Fig. 27. The superimpositions of potentials are for the eight consecutive nodes from the 7th to the 14th as shown in the first row, except in the second, third, fourth and fifth rows where the potentials shown are at the 7th, 8th, 9th and 10th nodes. The *asterisk* (in the columns) indicates the demyelinated internodes between the given nodes. The extracellular potentials are given for a radial distance r = 1.0 mm.

decreases until the potential dies out. However, the potential shapes in the regions distal to the demyelinated zone are normal. The potentials show repetitive discharges in the ALS cases (Fig. 27, third panel). Their amplitudes are smaller than the first one, and the intervals between the spontaneous discharges decrease until the repetitive activity dies out.

The extracellular potentials in the PSD and PISD cases show increased polyphasia, which is a characteristic feature of the compound motor action potentials (CMAPs) in patients with uniform demyelinating neuropathies (Donofrio and Albers 1990). The extracellular potentials show repetitive discharges in the three ALS cases. These discharges are similar to those of the CMAPs in patients with ALS (Finsterer et al. 1997, Rowinska-Marcinska et al. 1997). We speculate that abnormal CMAPs in demyelinating neuropathies and ALS subtypes are probably based on abnormalities obtained in the extracellular potentials of corresponding single motor nerve fibres. The action potentials, generated by long-lasting suprathreshold depolarizing current stimuli and their corresponding extracellular potentials in the ISD or IFD cases, are consistent with the effect of the uniform (ISD) or restricted (IFD) reduction of the myelin lamellae along the fibre length. And they are characteristic for simulated CMT1A and GBS (see Table 1 on page 34). The same potentials in the PSD and PFD cases are consistent with the effect of the uniform (PSD) or restricted (PFD) reduction of the paranodal seal resistance along the fibre length, and are characteristic for simulated CIDP and MMN, respectively. In the PISD and PIFD cases, action potentials and their corresponding extracellular potentials are consistent with the effect of the reduced myelin lamellae, additionally increased by reduced paranodal seal resistance along the fibre length. These potentials are characteristic for demyelinations, which are heterogeneous, such as in some simulated CIDP subtypes and MMN.

The repetitive firing in the action potentials generated by long-duration suprathreshold depolarizing current stimuli and the repetitive firing in their corresponding extracellular potentials in the three ALS cases, are consistent with the repetitive activation and reactivation of the sodium channels (Na^+) in the nodal and internodal axolemma in these cases. The resulting axonal superexcitability is based on the effect of the blocked potassium channels in the nodal or internodal axolemma along the fibre length. In brief: (i) the nodal fast potassium (K_f^+) channels are blocked in the ALS1 case; (ii) the nodal slow (K_s^+) and internodal fast (K_f^+) potassium channels are blocked in the ALS2 case, and; (iii) the blocked fast (K_f^+) and slow (K_s^+) potassium channels beneath the myelin sheath are in combination with the blocked nodal slow (K_s^+) potassium channels in the ALS3 case. According to us, this axonal superexcitability in response to long-duration suprathreshold depolarizing current stimuli is a prelude to cell (neuron) death. The results presented here are based on our previous studies (Stephanova and Daskalova 2002, 2005a,b, 2008, Stephanova and Alexandrov 2006, Stephanova et al. 2005, 2006a,b, 2007a, Stephanova 2006, 2010).

Abnormalities in the Strength-Duration Time Constants, Rheobasic Currents and their Mechanisms

Strength-duration and charge-duration curves for the normal, mild ISD, PSD, PISD (Fig. 29, first row), severe IFD, PFD, PIFD (Fig. 29, second row), ALS1, ALS2 and ALS3 (Fig. 29, third row) cases of human motor nerve fibres are shown. Histograms are also used to provide a better illustration of strength-duration time constants and rheobasic currents (Fig. 30). In the different strength-duration curves, the threshold currents are significantly higher in the systematically demyelinated cases (ISD, PSD and PISD, in Fig. 29—first row on the left) than in the normal one. Note that the strength-duration curves in the normal and abnormal cases are not natural exponential expressions, and their corresponding charge-duration curves are not straight lines (ISD, PSD and PISD, in Fig. 29—first row on the right). The strength-duration time constant is substantially longer for the ISD case and substantially shorter for the PSD case than for the normal one (Fig. 30—first row on the left). There is an inverse relationship between the strength-duration time constants and rheobasic currents for the normal, PSD and PISD cases, but this is not the same for the ISD case. In this case, the strength-duration time constant and rheobasic current are increased. The strength-duration time constants are 0.291, 0.583, 0.092, 0.211 ms and the rheobasic currents are 0.388, 0.568, 0.842, 1.020 nA for the normal, ISD, PSD and PISD cases, respectively.

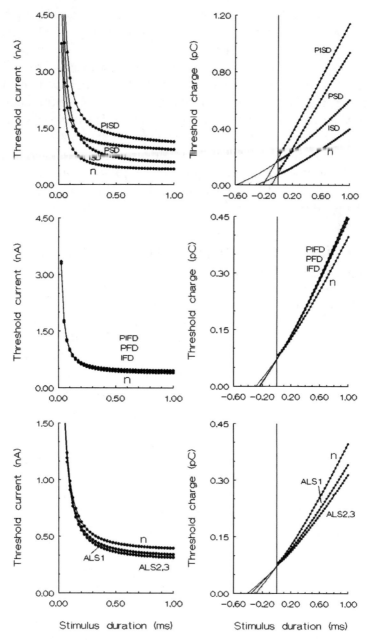

Fig. 29. Comparison between the strength-duration curves (left hand column) and charge-duration curves (right hand column) of human motor nerve fibres in the normal (n), mild ISD, PSD, PISD, severe IFD, PFD, PIFD and ALS1, ALS2, ALS3 cases.

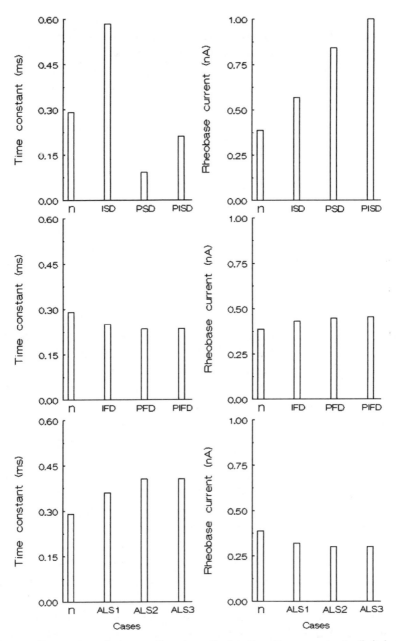

Fig. 30. Comparison between the strength-duration time constants (left hand column) and rheobasic currents (right hand column) of human motor nerve fibres in the normal (n), mild ISD, PSD, PISD, severe IFD, PFD, PIFD and ALS1, ALS2, ALS3 cases.

The progressively greater increase in focal loss of myelin slightly increases threshold currents in the cases of severe demyelination (IFD, PFD and PIFD in Fig. 29—second row on the left). They are slightly higher in the IFD, PFD and PIFD cases than in the normal one. The strength-duration curves for the severely focally demyelinated cases again are not natural exponential expressions, and their corresponding charge-duration curves are not straight lines (IFD, PFD and PIFD, in Fig. 29—second row on the right) The strength-duration time constants are slightly shorter than the normal one (Fig. 30—second row on the left). There is an inverse relationship between the strength-duration time constants and rheobasic currents for the normal and abnormal cases. The strength-duration time constants are 0.291, 0.250, 0.236 and 0.235 ms for the normal, IFD, PFD, PIFD cases, respectively. The rheobasic currents are 0.388, 0.431, 0.447 and 0.454 nA, respectively, for the same cases.

Strength-duration and charge-duration curves for the normal and abnormal three ALS subtypes are also shown (Fig. 29—third row). In the different strength-duration curves, the threshold currents are lower in the ALS1 and ALS2 cases than in the normal case. No differences are obtained between the strength-duration curves, charge-duration curves, strength-duration time constants and rheobasic currents in the ALS2 and ALS3 cases. Strength-duration curves for the three ALS cases again are not natural exponential expressions, and their corresponding charge-duration curves are not straight lines (ALS1, ALS2 and ALS3, in Fig. 29—third row on the right). Strength-duration time constants are longer for the ALS1 and ALS2 cases than the normal case. Strength-duration time constants increase and the rheobasic currents decrease with each successive ALS subtype (Fig. 30—third row). The strength-duration time constants are 0.291, 0.361, 0.406 and 0.406 ms for the normal, ALS1, ALS2, and ALS3 cases, respectively. The rheobasic currents are 0.388, 0.321, 0.301 and 0.301 nA, respectively, for the same cases.

These results show that the strength-duration time constants are longer in the ISD case and shorter in the PSD and PISD cases than the normal one. However, they are slightly shorter in the IFD, PFD and PIFD cases. Longer and shorter strength-duration time constants are characteristic for CMT1A patients (Nodera et al.

2004, Nodera and Kaji 2006) and CIDP patients (Cappelen-Smith et al. 2001, Nodera and Kaji 2006), respectively. According to the classical theory, the strength-duration time constant (chronaxie) of an excitable structure is based on the single passive (RC) membrane parameters in parallel and can be calculated by the classical Weiss's formula (Weiss 1901) applied to the linear charge-duration curve. In the human motor nerve fibre, RC parameters in parallel are placed in the nodal and internodal axolemma as well as in the myelin sheath. Because of this complicated situation, the strength-duration curves of normal and demyelinated fibres are not natural exponential expressions, and their corresponding charge-duration curves are not linear. This claim is supported by the strength-duration time constant obtained in the ISD case, where the passive myelin parameters (R_{my}, C_{my} in parallel) are uniformly changed along the fibre length (see Fig. 1 on page 19). The strength-duration time constant obtained in the PSD case depends on the reduced paranodal seal resistance (R_{pn}) along the fibre length. On the other hand the time constant obtained in the PISD case depends on the simultaneously changed passive parameters (R_{my}, C_{my} in parallel) in the myelin sheath and the reduced resistance (R_{pn}) in the paranodal segments (see Table 1 on page 34), where the changes of these parameters are uniform along the fibre length. The changes in the strength-duration time constants obtained for the simulated focally demyelinated cases (IFD, PFD and PIFD) are small. The time constants are near normal as the changed membrane parameters mentioned above are restricted to only three internodal segments along the fibre length. Such near normal strength-duration time constants are characteristic for GBS patients with acute motor axonal neuropathy and acute inflammatory demyelinating polyneuropathy (Kuwabara et al. 2002).

The strength-duration and charge-duration curves, strength-duration time constants and rheobasic currents are abnormal in the ALS1, ALS2 and ALS3 cases. The equal time constants and rheobasic currents obtained in the ALS2 and ALS3 cases show that the blockage of the internodal slow potassium channels does not contribute to these parameters. In the abnormal cases, the strength-duration time constants are longer than the normal one. Such time constants are typical for ALS patients (Mogyoros et al. 1998).

According to us and to some other authors (Bostock and Rothwell 1997), the strength-duration time constant may also be regarded as an axonal property. The increased strength-duration time constant in the ALS1 case may be explained by the increase passive and changed active axonal parameter values (see Table 2 on page 36). In the ALS2 case, the imbalance between the nodal and internodal slow and fast potassium (K_s^+, K_f^+) channels may cause an additional increase in the time constant.

According to the classical theory, there is a reciprocal relationship between the strength-duration time constant and rheobasic current in healthy axons, i.e., when the time constant increases, the rheobasic current decreases and vice versa. Such a reciprocal relationship between the strength-duration time constant and rheobasic current is obtained for the simulated demyelinating neuropathies and ALS, with the exception of the simulated CMT1A, where the strength-duration time constant and rheobasic current are increased. The rheobasic currents are lower in the simulated ALS subtypes and are higher in the simulated demyelinating neuropathies (CMT1A, CIDP, CIDP subtypes, GBS and MMN) (Fig. 30—right hand column). These changes in rheobasic currents can be attributed to the lower vs. higher threshold currents in the corresponding cases, as can be seen from the strength-duration curves for these cases (Fig. 29).

The normal strength-duration time constant of 0.291 ms, that we calculated is in the clinical range (0.283–0.629 ms) measured at 50% CMAP (compound motor action potential) maximal amplitude in healthy controls (Cappelen-Smith et al. 2001). The closest to our value is the one measured by Panizza et al. (1994, 1998). It was previously shown by us that: (i) the strength-duration time constants progressively increase with the progressive increase of demyelination in simulated CMT1A (Stephanova et al. 2005) and (ii) the strength-duration time constants progressively decrease with the progressive increase of demyelination in simulated CIDP and CIDP subtypes (Stephanova and Daskalova 2005a,b). These findings are also clinically proven. The clinically obtained ranges for the strength-duration time constants are usually determined for skin temperature slightly above 32°C. According to Keirnan et al. (2001) these ranges shift to lower values for skin temperature of 37°C. The strength-duration time constants also depend on the age of the patients (Mogyoros et al. 1998, Nodera et al. 2004) etc. Consequently, time constants depend on a number of factors.

Abnormalities in the Recovery Cycles and their Mechanisms

The axonal excitability changes in the normal, demyelinated (systematically, focally) and ALS subtypes during 100 ms recovery cycles are illustrated in Fig. 31a–c. The recovery cycles show that excitability changes in the normal and ISD cases are similar (Fig. 31a). Axonal types are initially unexcitable, then excitable with a raised threshold and after about 2.5 ms they are superexcitable. In the unexcitable case, axons are in the absolute refractory period, during which they cannot generate a second potential no matter how strong the testing depolarizing stimuli are. Then the axons are in the relative refractory period, during which a stronger than conditioning stimulus is required to generate a second potential. In the superexcitable case, the testing stimulus generating the second action potential is less than the conditioning stimulus. Axonal superexcitability is usually followed by late subexcitability. The ISD axon has greater refractoriness, superexcitability and less late subexcitability than the normal axon. There is an increase in the refractoriness without an increase in the relative refractory period in this demyelinated axon. Recovery cycles show that axonal excitability changes in the PSD and PISD cases are similar. Both axonal subtypes have abnormally greater superexcitabilities. In these cases, the recovery cycles are without relative refractory periods and have only superexcitable and subexcitable periods, compared to normal one.

The same investigations of axonal excitability changes during 100 ms recovery cycles are repeated for the IFD, PFD and PIFD cases of the fibres and are illustrated in Fig. 31b. The recovery cycles show that axonal excitability changes in all focally demyelinated cases are similar. Axons have slightly less refractoriness (the increase in threshold current during the relative refractory period), greater superexcitability and less late subexcitability than the normal axon. However, the recovery cycles are near-overlap for the PFD and PIFD cases. The excitability changes of the human motor axons in the normal and ALS cases during the 100 ms recovery cycles are also illustrated (Fig. 31c). The recovery cycles show

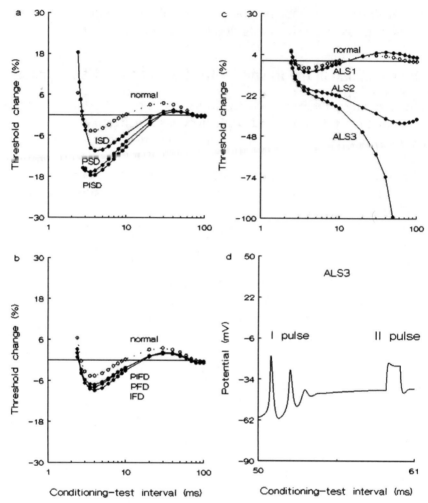

Fig. 31. Comparison between the recovery cycles in the normal (*dotted lines*), mild ISD, PSD, PISD **(a)**, severe IFD, PFD, PIFD **(b)**, ALS1, ALS2, and ALS3 **(c)** cases of human motor nerve fibres. For all cases, the y-axis is defined as 100 x ($I_{test} - I_{cond}$)/I_{cond} (%), where I_{test} (nA) and I_{cond} (nA) are the threshold currents of the testing and conditioning pulses, respectively. The temporal distribution of the action potentials as a function of the conditioning—test interval in the ALS3 case **(d)**. The conditioning and testing current stimuli are with a duration of 1 ms.

that the excitability changes of the normal and abnormal axons are considerably different. The ALS1 axon has less refractoriness, greater superexcitability and late subexcitability than the normal one. For the ALS2 case, the superexcitability increases markedly and the axon remains superexcitable to the end of the investigated period. For the ALS3 case, the superexcitability increases rapidly and abnormalities are obtained in the recovery cycle beyond 50 ms conditioning-test intervals. In these cases, each second stimulus applied to the internodal axolemma is blocked (Fig. 31d) and this is a result of spontaneous axonal activities caused by the first action potential.

The axonal excitability recovery cycle depends on the regenerative membrane depolarization caused by the first (conditioning) afterpotential and could be explained by the delay-dependent second (testing) potential. During the absolute refractory period, when the inactivation of the sodium (Na^+) channels is low and the activation of the potassium (K^+) channels is high, the threshold potential (Vt) and the corresponding threshold current for the testing pulse are increased due to the increase of the critical membrane (Ec) potential [(Vt = Ec-Er); Er, resting membrane potential]. The critical membrane potential decreases with the increase of the conditioning-test interval. Since the membrane potential remains above the resting potential (Er), the threshold potential (Vt) becomes lower than that of the initial one. The result is a change from subnormality to supernormality of the axonal excitability. Consequently, in the recovery cycle, the axon has superexcitability when the threshold current of the testing stimulus is less than the threshold current of the conditioning stimulus and vice versa. The axon has subexcitability when the testing stimulus necessary to generate the second potential is stronger than the conditioning one.

The pattern of recovery cycles obtained in the ISD, PSD and PISD cases is characteristic for patients with CMT1A, CIDP and CIDP subtypes, respectively (Nodera et al. 2004, Sung et al. 2004), whereas the recovery cycles obtained in IFD, PFD and PIFD are characteristic for the median motor nerves in AIDP patients (Kuwabara et al. 2002, Nodera and Kaji 2006).

The pattern of recovery cycles in the simulated ALS1 and normal cases is similar. However, the refractoriness is less and the superexcitability is greater than those in the normal case, and such results are also discussed by Mogyoros et al. (1998) for ALS1 patients and normal controls. Vastly greater superexcitabilities in the ALS2 case and abnormalities obtained in the ALS3 case show how the recovery cycles are altered with the increase of the axonal ion channel dysfunction. The abnormally increased axonal excitability in the ALS3 case leads to repetitive discharges, generated by the first action potential and to blockage of each second stimulus, applied to the internodal axolemma along the fibre length (Fig. 31d). The superexcitability of the nodal and internodal axolemma, caused by the continuously activated and reactivated sodium channels in these compartments, can be regarded as a prelude to neuron death and probably as a reason for motor neuron degeneration in this disease.

The simulations, presented in this chapter, were done using a well validated model. It is possible to reproduce a wide range of experimental data on the excitability properties in human demyelinating neuropathies and ALS. This model provides an objective interpretation of the mechanisms of the excitability abnormalities obtained in simulated hereditary, chronic, acquired demyelinating neuropathies and in the three simulated ALS subtypes, which until now have not been sufficiently understood.

In summary, from axonal excitability measurements and changes in excitability produced by nerve impulses, it is possible to infer information about the electrophysiological properties of peripheral and central nerves. Studies on threshold electrotonus and excitability indices recorded from stimulated motor nerves in human demyelinating neuropathies such as CMT1A, CIDP, CIDP subtypes, GBS, MMN and motor neuron diseases, such as ALS show that these excitability properties are not identical.

The investigations presented here on the potentials (action, electrotonic, extracellular) and excitability indices (strength-duration time constants, rheobasic currents, recovery cycles) of the three simulated systematic, focal demyelinations and three progressively greater degrees of uniform axonal dysfunctions along the length of human motor nerve axons, confirm that their excitability properties are not identical and can be used as specific indicators for these disorders. The results show that the changes obtained in simulations replicate those recorded in patients with demyelinating neuropathies and ALS. The analysis of the excitability properties show that: (i) the mild ISD is a specific indicator for CMT1A; (ii) the mild PSD and PISD are specific indicators for CIDP and its subtypes; (iii) the severe IFD and PFD, PIFD are specific indicators for acquired demyelinating neuropathies such as GBS and MMN; and (iv) the simulated progressively greater degrees of axonal dysfunctions in the ALS1, ALS2 and ALS3 cases are specific indicators for ALS Type1, Tape2 and Type3 motor neuron disease.

In the mild ISD case the following are specific indicators for simulated CMT1A: 1) the slowed conduction velocity, 2) increased hyperpolarizing electrotonic response to 100 ms stimuli, 3) homogeneous membrane polarization in the case of accommodation, 4) prolonged strength-duration time constant, 5) increased rheobasic current and 6) increased superexcitability period of the axonal excitability, recovered at the end of the investigated 100 ms cycle. In the mild PSD and PISD cases the following are specific indicators for simulated CIDP and CIDP subtypes: 1) the slowed conduction velocities, 2) abnormally increased hyperpolarizing responses to 100 ms current stimuli, 3) homogeneous membrane polarization in the case of accommodation, 4) increased polyphasia in the extracellular potentials, 5) shortened strength-duration time constants, 6) increased rheobasic currents and 7) abnormally increased superexcitability periods of the axonal excitability, recovered at the end of the investigated 100 ms cycles.

In the severe IFD and PFD, PIFD cases the following are specific indicators for simulated GBS and MMN: 1) the blockage of action potential propagation, 2) small drop to a minimal amplitude within the demyelinated zone, and small rise to a maximum amplitude outside this zone in the electrotonic potentials, 3) heterogeneous membrane polarization 4) dying out of the extracellular potentials within the demyelinated zone, and normal extracellular potentials without this zone, 5) slightly decreased strength-duration time constants, 6) slightly increased rheobasic currents and 7) superexcitability periods of axonal excitability, recovered at the end of the investigated 100 ms cycles.

In the ALS1 case, the following are specific indicators for simulated ALS Type1: 1) the repetitive discharges in the extracellular potential, 2) prolonged strength-duration time constant, 3) decreased rheobase and 4) slightly increased superexcitability period of the axonal excitability, recovered at the end of the investigated 100 ms cycle. In the ALS2 case, the following are specific indicators for simulated ALS Type2: 1) the repetitive discharges in the early part of the depolarizing electrotonic potential as well as in the extracellular potential, 2) prolonged strength-duration time constant, 3) decreased rheobase and 4) abnormally increased superexcitability period of the axonal excitability, that is not recovered by the end of the investigated 100 ms cycle. In the ALS2 case each third stimulus (if it is applied) will be blocked as a result of the repetitive firing caused by the previous second stimulus. In the ALS3 case, the following are specific indicators for simulated ALS Type3: 1) the repetitive discharges in the early part of the depolarizing electrotonic potential as well as in the late part (100–200 ms) of the hyperpolarizing electrotonic potential and in the extracellular potential, 2) prolonged strength-duration time constant, 3) decreased rheobasic current and 4) abnormally increased axonal excitability, which leads to blockage of each applied second current stimulus in the investigated 100 ms conditioning-test interval.

The mechanisms defining the abnormalities obtained in the simulated demyelinating neuropathies and ALS were discussed in detail in this chapter. They show that the excitability property abnormalities in simulated demyelinating neuropathies are quite different from those in ALS because of the different fibre electrogenesis. The excitability abnormalities obtained in the simulated demyelinating neuropathies are based on the myelin sheath dysfunctions, whereas the excitability abnormalities obtained in the simulated ALS subtypes are based on the axonal dysfunctions along the length of human motor nerve fibres. These results show that the transition from conduction slowing (mild demyelinations) to conduction block (severe demyelinations) of action potential leads to amplification of the degree of the excitability property changes, as the direction of these changes is maintained. The results also show that the action potential propagation slowing is larger in simulated hereditary, chronic and acquired demyelinating neuropathies, than this in simulated ALS. Conversely, abnormalities in the accommodative and adaptive processes are larger in the ALS2 and ALS3 subtypes than in demyelinating neuropathies. With the increase of the axonal dysfunction from ALS1 to ALS3, the axonal excitability increases rapidly, which leads to repetitive firing in the electrotonic potentials as well as in the action and extracellular potentials in the case of adaptation of human motor nerve fibres. Consequently, simulated ALS axons are superexcitable when they are stimulated by long-duration subthreshold or suprathreshold stimuli. This axonal superexcitability leads to blockage of each third applied stimulus in the case of ALS2 and to blockage of each second applied stimulus in the case of ALS3 as a result of the repetitive firing caused by the previous applied stimulus. Superexcitability of the nodal and internodal axolemma, based on the continuous activation and reactivation of the ionic (mainly Na^+) channels in these compartments, can be regarded as a prelude to neuron death. And this is probably the reason for motor neuron degeneration in this disease.

Chapter IV

Effect of Myelin Sheath Aqueous Layers on the Excitability Properties of Simulated Hereditary and Chronic Demyelinating Neuropathies

Simulation of CMT1A, CIDP and CIDP Subtypes with Aqueous Layers within the Myelin Sheath

The myelin sheath is normally regarded as an electrical insulator. However, the typical picture which we have from electron microscopy of fixed myelin sheath is rather misleading. Studies of native myelin by X-ray and neutron diffraction have demonstrated that the myelin lamellae are not tightly compacted, but separated by cytoplasmic and extracellular spaces of about 4–5 nm, respectively, so that up to half the volume fraction of myelin is taken up by water (Kirschner and Caspar 1972, Kirschner et al. 1984). Moreover, the water in both aqueous layers contains salts that can be washed out or exchanged with the extracellular solution. The calcium ions in the extracellular compartment have a special role in maintaining myelin structure by electroctatic interaction with fixed negative charges on the membrane (Ropte et al. 1990), but other ions can be shown to diffuse readily throughout both aqueous compartments (Blaurock 1971). Consequently, the myelin sheath, taking into account its aqueous layers, can be regarded as an electrical conductor. The hypothesis that the aqueous layers endow the myelin sheath with longitudinal conductance much greater than its radial conductance was previously demonstrated by our multi-layered myelin sheath model of the human motor nerve fibre (Stephanova 2001). In this study, it was found that conduction velocity of the action potential depends on the longitudinally conducting aqueous layers which assist action potential propagation. The conduction velocity was appreciably faster (by 8.6%).

Simulations of normal case and mild systematic demyelinations, each of them internodal, paranodal and paranodal internodal, were shown and discussed in Chapter III. Although the multi-layered model was used for the simulations, the complex structure of the myelin sheath was not taken into account. Simulations of the normal case and systematic demyelinations, each of them internodal, paranodal and paranodal internodal, are presented and discussed below (Fig. 32) using the same model that takes into account the aqueous layers within the myelin sheath. Three progressively greater degrees (two mild -70, 80% and one severe) of each systematic demyelination (i.e., ISDs, PSDs and PISDs for s=1, 2, 3) are investigated without versus with taking into account the aqueous layers within the myelin sheath. The 70% (ISD1, PSD1 and PISD1) and 80% (ISD2, PSD2 and PISD2) values of the myelin lamellae that are uniformly reduced along the fibre length are not sufficient to develop conduction block of the intracellular action potential. The reduction values 93% (ISD3)/93% (ISD3); 90% (PSD3)/89% (PSD4); and 82% (PISD3)/82% (PISD3) are the first degrees for achieving conduction block in the given demyelinated cases with vs. without taking into account the aqueous layers within the myelin sheath. The comparison between the multiple axonal excitability properties (such as action and electrotonic potentials, strength-duration time constants, rheobasic currents) without and with taking into account the aqueous layers within the demyelinated sheath, reveals the effect of the aqueous layers on the investigated axonal excitability properties.

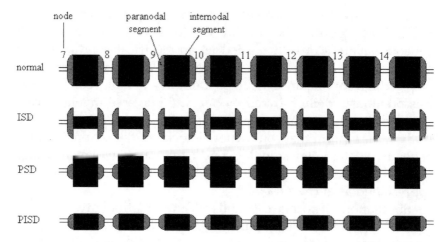

Fig. 32. Diagram of human motor nerve fibres from the 7th to the 14th nodes in the normal, ISD, PSD and PISD cases. Three progressively greater degrees (two mild –70, 80% and one severe) of each of the systematic demyelination (i.e., ISD1, ISD2, ISD3; PSD1, PSD2, PSD3; PISD1, PISD2 and PISD3) are simulated without vs. with myelin sheath aqueous layers.

The characteristic parameter values defining the investigated cases are given in Table 3, for the reader's convenience.

The 70% and 80% reduction values of the myelin lamellae are not arbitrary taken. It was previously found that the direction of the longitudinal current flow is the same as in the axoplasm from the 2nd to the 25th aqueous layer, whereas to the 50th lamella this direction is reversed and above the 75th lamella the longitudinal myelin currents are negligible (Stephanova 2001). The myelin reduction value for achieving conduction block depends on the type of demyelinations of a given subtype. It ranges from 93% to 82% for ISD3 and PISD3, respectively. These reduction values are the same for the ISD3 and PISD3 subtypes, without and with myelin aqueous layers. However, these values are different for the paranodal systematic demyelinations. Note the 90% myelin reduction value for the PSD3 case with aqueous layers and the 89% myelin reduction value for the PSD4 case without aqueous layers (see also Table 3). The results presented below are consistent with the interpretation that: (i) genetic factors causing changes in the internodal segments of the myelin sheath, without and with aqueous layers; (ii) immunological factors causing changes in

Table 3. Membrane parameter values characteristic for human motor nerve fibres in the normal and systematically demyelinated cases (ISDs, PSDs, PISDs for s=1, 2, 3), without and with taking into account the myelin sheath aqueous layers, when demyelinations are mild and severe. The PISDs are composed of ISDs and PSDs. N (number of myelin lamellae); R_{my} (myelin resistance); C_{my} (myelin capacitance); R_{aql} (longitudinal myelin resistance); R_{aqr} (radial myelin resistance); R_{pa} (periaxonal resistance); R_{pn} (paranodal seal resistance).

	Without aqueous layers	With aqueous layers
Normal		
N [lamellae]	150	150
R_{my} [MΩ]	250.0	250.0
C_{my} [pF]	1.5	1.5
R_{aql} [MΩ]	∞	21.3
R_{aqr} [MΩ]	∞	436.8
R_{pa} [MΩ]	300.0	1250.0
R_{pn} [MΩ]	125.0	140.0
PISD1 [70%]		
N [lamellae] **(ISD1)**	45	45
R_{pn} [MΩ] **(PSD1)**	37.5	42.0
PISD2 [80%]		
N [lamellae] **(ISD2)**	30	30
R_{pn} [MΩ] **(PSD2)**	25.0	28.0
PISD3 [82%]		
N [lamellae]	27	27
R_{pn} [MΩ]	22.5	25.2
ISD3 [93%]		
N [lamellae]	11	11
PSD3 [90%]		
R_{pn} [MΩ]	12.5	14.0
PSD4 [89%]		
R_{pn} [MΩ]	13.75	15.4

the paranodal segments of the myelin sheath, without and with aqueous layers; and (iii) simultaneously both of these factors causing changes in the myelin sheath, without and with aqueous layers, could be responsible for axonal excitability abnormalities obtained in simulated hereditary (CMT1A) and chronic (CIDP, CIDP subtypes) demyelinating neuropathies, respectively.

Effect of Myelin Sheath Aqueous Layers on the Potentials

Comparison of the action potentials for the normal (first row), ISD (a), PSD (b) and PISD (c) cases, without (*dotted lines*) and with (*continuous lines*) aqueous layers within the myelin sheath, is shown for human motor nerve fibres (Fig. 33). Potentials are presented at the 10th node only. The potential maxima at the 10th nodes are 38/38, 18/17, 11/9 mV in the normal, ISD1 and ISD2 cases without versus with aqueous layers, respectively (Fig. 33a). A decrease of conduction velocities is obtained for the abnormal cases, except for the normal case where aqueous layers increase the conduction velocity by 8.6%. Normal conduction velocity is 58 m/s without aqueous layers. Conduction velocities calculated from the times of potential maxima at the nodes are 31 and 23 m/s for the ISD1 and ISD2 cases without aqueous layers, respectively. Velocity remains almost unchanged (30.6 m/s) in the ISD1 case and decreases by 3.4% in the ISD2 case when aqueous layers are taken into account. The progressively greater increase in uniform loss of the myelin lamellae, without vs. with their aqueous layers, blocks the invasion of potentials in the ISD3 case. Thus, with the increase of the demyelination from ISD1 to ISD3, conduction failure occurs rapidly. The reduction value of myelin lamellae for achieving conduction block is the same for both cases without and with aqueous layers within the myelin sheath (see Table 3).

The maximal amplitudes are 21/23, 12/13 mV in the PSD1 and PSD2 cases without vs. with aqueous layers, respectively (Fig. 32b). Conduction velocities calculated from the times of potential maxima at the nodes are 41 and 35 m/s for the PSD1 and PSD2 cases without aqueous layers, respectively. For the PSD1 and PSD2 cases, the addition of the aqueous layers increases the conduction velocities by 6.9% and 8.6% respectively. The progressively greater increase in uniform reduction of the paranodal seal resistance blocks the invasion of potentials in the severe PSD3 case, when the aqueous layers are taken into account. Significantly, the aqueous layers in the 89% demyelinated case (PSD4) restore action potential

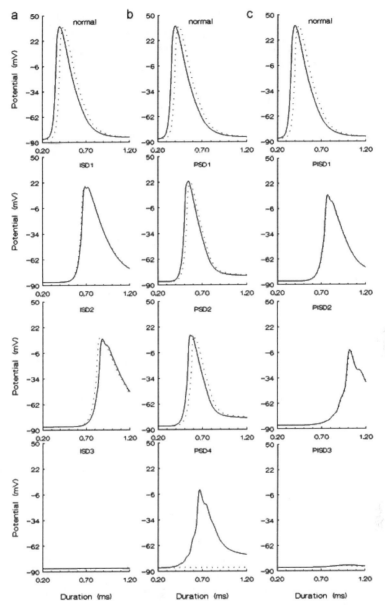

Fig. 33. Comparison between the action potentials of human motor nerve fibres in the normal, ISD **(a)**, PSD **(b)** and PISD **(c)** cases without aqueous layers presented by *dotted lines* and those with aqueous layers presented by *continuous lines*. The potentials are given at 10th node only. The data show that aqueous layers in the 89% demyelinated case (PSD4) restore action potential propagation, which is initially blocked when aqueous layers are not taken into account.

propagation (Fig. 33b, *continuous line*—in the bottom), which was initially blocked when the aqueous layers were not taken into account (Fig. 33b, *dotted line* in the bottom). These data show that aqueous layers markedly promote action potential propagation, normally and in cases of up to 89% paranodal demyelination. The maximal amplitude of the restored potential is −1.5 mV and its conduction velocity is 19 m/s.

An overlap of the action potentials at the nodes is obtained for all PISD cases, without and with aqueous layers within the myelin sheath (Fig. 33c). The maximal amplitudes of the potentials are 9 and −3 mV for the PISD1 and PISD2 cases, respectively. Conduction velocities are 25 and 13 m/s for the PISD1 and PISD2 cases, respectively.

Comparison of the electrotonic potentials for the normal (first row), ISD (a), PSD (b) and PISD (c) cases, without (*dotted lines*) and with (*continuous lines*) myelin aqueous layers, is also shown for human motor nerve fibres (Fig. 34). Potentials in response to 100 ms applied polarizing current stimuli (±40% of threshold) are presented at the 10th nodes only. During and after these applied stimuli there are no differences in potentials between the normal case (first row in Fig. 34) and all PISD cases (Fig. 34c) without vs. with aqueous layers within the myelin sheath. There are almost no differences in the depolarizing electrotonic potentials between the normal and remaining ISD and PSD cases without vs. with aqueous layers (Fig. 34a,b). When comparing the normal case with the ISD1–ISD3 and PSD1–PSD3 cases without aqueous layers, the resulting differences are the abnormally increase in the early and late parts of the hyperpolarizing responses (Fig. 34a,b). For the ISD1, ISD2 and ISD3 cases an additional increase in the early and late parts of these responses is obtained when the aqueous layers are taken into account (Fig. 34a) and electrotonic hyperpolarizing potentials Ehs (100 ms) are increased by 3, 5 and 6%, respectively. However, for the PSD1, PSD2 and PSD3 cases, a further decrease in the early and late parts of these responses is observed when the myelin aqueous layers are taken into account and the electrotonic hyperpolarizing potentials Ehs (100 ms) are decreased by 2.8, 2.5 and 2.3%, respectively.

When the normal case is compared to the more severe PISD2 and PISD3 cases, the resulting differences are the greater increase

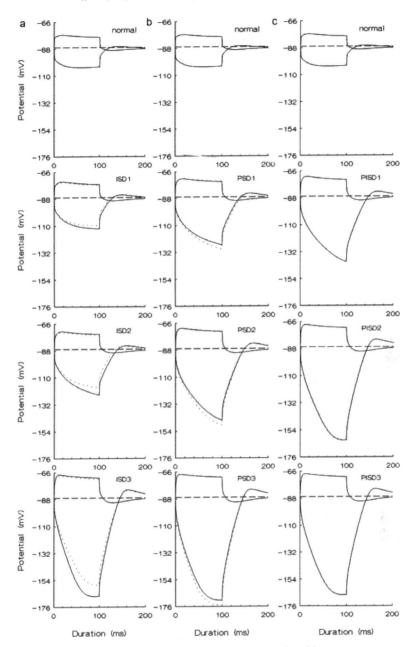

Fig. 34. Comparison between the electrotonic potentials of human motor nerve fibres in the normal, ISD **(a)**, PSD **(b)** and PISD **(c)** cases without aqueous layers presented by *dotted lines* and those with aqueous layers presented by *continuous lines*. The *dashed lines* indicate the resting potential.

in the early part of the depolarizing responses and the abnormally greater increase in the early and late parts of the hyperpolarizing responses (Fig. 34c). Potentials are mostly abnormal in the PISD3 case.

The results presented here show that aqueous layers within the myelin sheath modulate potentials (action and electronic) of human motor nerve fibres when systematic demyelinations are internodal and paranodal. The loss of myelin lamellae and their corresponding aqueous layers in the ISD1 and ISD2 cases additionally slows the conduction velocity of action potentials in comparison to cases when aqueous layers are not taken into account. Conduction velocities are in the range of 31/30.6–23/22.2 m/s for the cases without vs. with myelin aqueous layers, respectively. The mean median nerve conduction velocities, measured in patients with CMT1A, are 21±5.7 m/s (Hattori et al. 2003) and 16.5 m/s in the range of 5–35 m/s (Carvalho et al. 2005, Marques et al. 2005). The relatively weak effect of the aqueous layers on the propagating processes in the investigated ISD1 and ISD2 cases can account for the same conduction block reduction values (93%/93%) of the myelin lamellae without vs. with aqueous layers.

The uniform reduction of the paranodal seal resistance in the PSD1 and PSD2 cases, taking into account aqueous layers within the myelin sheath, additionally increases the conduction velocity of the action potentials in comparison to cases when aqueous layers are not taken into account. Calculated conduction velocities are in the range of 41/44–35/38 m/s for the cases without vs. with aqueous layers, respectively. Conduction velocities in patients with CIDP, measured in median motor axons in the elbow-wrist segment, are in the range of 54–20 m/s (Cappelen-Smith et al. 2001, Sung et al. 2004). For the one severely demyelinated case, for which propagation without aqueous layers was blocked, the aqueous layers restore the action potential propagation. In the case just befor the blockage (PSD4 –89%), the conduction velocity of action potential is 19 m/s.

The results presented here also show that aqueous layers within the myelin sheath do not modulate action and electrotonic potentials of human motor fibres when the systematic demyelinations are simultaneously paranodal internodal. This result is quite unexpected because a significant effect of the aqueous layers on the action and electrotonic potentials is obtained when demyelinations are internodal systematic or paranodal systematic. Changes in potentials obtained in the PISD cases are in good accordance with the data from patients in the CIDP subgroups (Sung et al. 2004).

Action potential changes in the normal and abnormal ISD, PSD and PISD cases without or with aqueous layers within the myelin sheath are determined by the kinetics of the currents passing through the nodal and internodal axolemma. Mechanisms of the action potential conduction slowing based on the kinetics of the currents passing through the mild ISD and PISD cases without aqueous layers, were thoroughly described and discussed in Chapter III; and they are the same as in the cases in which the aqueous layers are also taken into account. Consequently, their detailed description and discussion will not be given again. The aqueous layers have a relatively weak effect on the propagating processes in the internodally systematically demyelinated cases (i.e., in the simulated cases of CMT1A). However, a significantly greater effect of the aqueous layers is obtained on the propagating processes in the paranodally systematically demyelinated cases (i.e., in the simulated cases of CIDP) as a result of the dysfunction of each paranodal segment along the fibre length. The mechanisms of the increased action potential propagation in the simulated CIDP cases, when aqueous layers are taken into account, will be discussed in more details below. In this case, the action potential generation at the nodal segments is again defined mainly by the sodium current (I_{Na}), as the contribution of fast and slow potassium currents (I_{Kf}, I_{Ks}) to the total nodal ionic currents (I_i) is negligible. The externally recorded nodal inward current (I_m) is again less than the current across the nodal axolemma (I_a), because of current flow through the paranodal seal resistance. Nevertheless, the paranodal seal resistances in the PSD1, PSD2 and PSD3 cases with aqueous layers $(R_{pn}, 42, 28$ and 15.4 MΩ) are higher than those without aqueous layers $(R_{pn}, 37.5, 25$ and 13.75 MΩ, respectively) and the periaxonal resistance $(R_{pa}, 1250$ MΩ) is considerably higher in the case with aqueous layers, the equivalent periaxonal resistance $[R_{pa}^*, 20.9$ MΩ (Stephanova 2001)] is lower than the aqueous-layered longitudinal resistance $(R_{aql}, 21.3$ MΩ) [see Table 3 on page 97]. The first conducting longitudinal aqueous layer (R_{aql1}) in parallel with the periaxonal layer (R_{pa}) defines the equivalent periaxonal layer (R_{pa}^*) [see Fig. 1 on page 19], which assists action potential propagation. The periaxonal resistance in the case without aqueous layers is higher (R_{pa}, 300 MΩ, Table 3) than the equivalent periaxonal resistance. Consequently, the decrease of the paranodal seal resistance in the cases with aqueous layers results in an edditional increase of the longitudinal current flow through the paranodal axolemma to the periaxonal space. As a result, conduction velocities of action potentials are additionally increased in the simulated CIDP cases.

A significantly higher effect of the aqueous layers is obtained on the accommodative processes rather than on the propagating processes, as a result of the almost complete dysfunction of the internodal axolemma along the fibre length in the investigated ISD cases. In comparison with the cases without aqueous layers, an additional increase of the hyperpolarizing electrotonic potentials in the ISD cases is demonstrated when aqueous layers are taken into account. In the PSD cases investigated, all internodal compartments with their aqueous layers are not changed, and this can explain the negligible and relatively weak effect of the aqueous layers on the depolarizing and hyperpolarizing electrotonic potentials.

The mechanisms of electrotonic potential changes, as well as the kinetics of the currents passing through the mild ISD, PSD and PISD cases without aqueous layers, were thoroughly described and discussed in Chapter III. Consequently, a detailed description and discussion will not be given again. In brief, during the depolarizing current pulses, the contribution of the nodal slow potassium outward current to both the nodal total ionic current and the generation of the electrotonic potential, is large and almost the same for the normal and demyelinated cases without vs. with aqueous layers. Much of the externally recorded nodal current does not flow across the nodal axolemma, but passes longitudinally via the paranodal seal resistance to the periaxonal space. During the hyperpolarizing current pulses, the contribution of the nodal total ionic currents in the normal and abnormal cases is negligible, whereas at the paranodal and internodal segments, the inward rectifying and leakage currents dominate in the total ionic currents. With the increasing loss of myelin lamellae and corresponding aqueous layers, current flow during the fast electrotonic phase (0–10 ms) increases quickly, and this results in an additional increase of the rectifying and leakage currents. These mechanisms explain the significant differences in the results obtained between the polarizing electrotonic potentials in the simulated CMT1A cases without vs. with aqueous layers.

With the decrease of the paranodal seal resistance, the current flow during the fast hyperpolarizing electrotonic phase (0–10 ms) increases quickly, and this results in an increase of the rectifying and leakage currents in the cases without aqueous layers. However, the paranodal seal resistances in the PSD cases with aqueous layers are higher than those without aqueous layers. As a result, a further reduction in the fast electrotonic phase is observed, reflecting the decreased rectifying and leakage currents in the simulated CIDP cases. The results show that the potential abnormalities obtained in the PSD cases are totally opposite to those obtained in the ISD cases, when the aqueous layers are taken into account. However, the myelin sheath aqueous layers do not modulate potentials when the systematic demyelinations are mixed, i.e., paranodal internodal (PISDs). Consequently, the reciprocally opposed effects of the aqueous layers on the excitability axonal properties are neutralized, when the demyelinations are heterogeneous, such as in some simulated CIDP subtypes.

Effect of Myelin Sheath Aqueous Layers on the Strength-Duration Time Constants, Rheobasic Currents and Recovery Cycles

To expand our studies on axonal properties, the effect of the myelin sheath aqueous layers on the strength-duration properties in simulated ISD, PSD and PISD cases is compared. The strength-duration curves (Fig. 35) and charge-duration curves (Fig. 36) in the normal (n), ISD (a), PSD (b) and PISD (c) cases of human motor nerve fibres without (Figs. 35 and 36—left hand columns) and with (Figs. 35 and 36—right hand columns) aqueous layers are shown. Histograms are also used to provide a better illustration of strength-duration time constants and rheobasic currents (Fig. 37). The threshold currents are significantly higher in all investigated systematically demyelinated cases without vs. with aqueous layers than in the normal ones in the different strength-

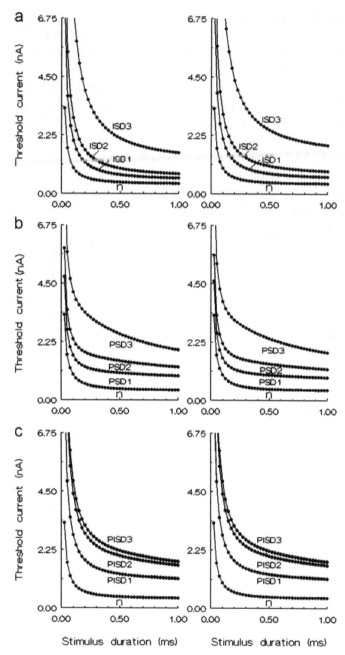

Fig. 35. Comparison between the strength-duration curves of human motor fibres in the normal (*n*), ISD **(a)**, PSD **(b)** and PISD **(c)** cases without (left column) and with (right column) aqueous layers within the myelin sheath.

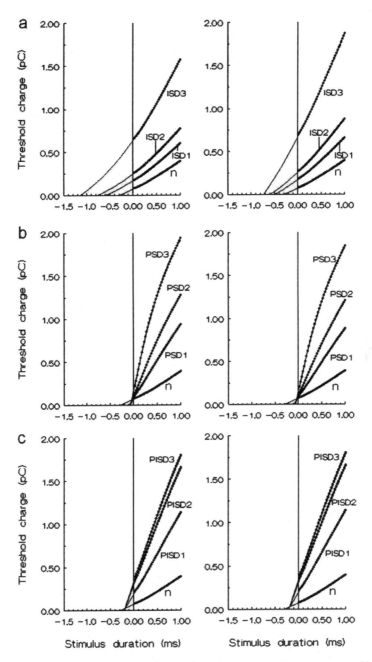

Fig. 36. Comparison between the charge-duration curves of human motor fibres in the normal (*n*), ISD (**a**), PSD (**b**) and PISD (**c**) cases without (left column) and with (right column) aqueous layers within the myelin sheath.

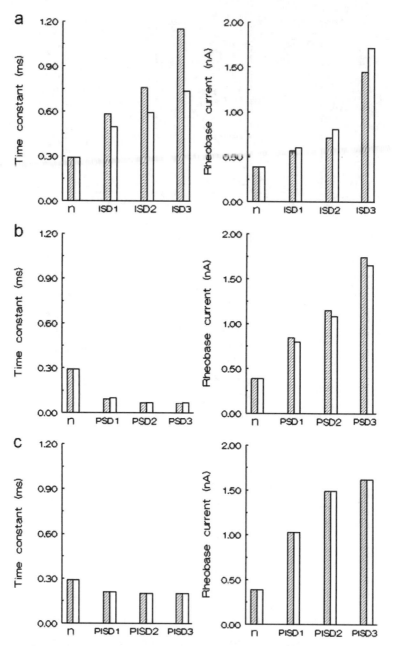

Fig. 37. Comparison between the strength-duration time constants and rheobasic currents of human motor fibres in the normal (*n*), ISD **(a)**, PSD **(b)** and PISD **(c)** cases without (*hatched bars*) and with (*unhatched bars*) aqueous layers within the myelin sheath.

duration curves. There is an inverse relationship between the strength-duration time constants (longer per shorter) which are 0.583/0.497, 0.757/0.592 and 1.149/0.736 ms for the ISD1, ISD2 and ISD3 cases without vs. with myelin aqueous layers, respectively. However, an elongation of the strength-duration time constant occurs more rapidly with the increase of the demyelination from ISD1 to ISD3 in the case without aqueous layers. The strength-duration time constant is substantially longer (1.149 ms) without aqueous layers and longer (0.736 ms) with aqueous layers than the normal one (0.291 ms) for the ISD3 case.

There is an inverse relationship between the strength-duration time constants, however, they are shorter per longer (0.093/0.099, 0.065/0.068 and 0.064/0.068 ms) for the PSD1, PSD2 and PSD3 cases without vs. with myelin aqueous layers, respectively. An overlap of the strength-duration time constants and rheobasic currents is obtained for all investigated PISD cases without vs. with aqueous layers. The strength-duration time constants are substantially shorter 0.211, 0.202 and 0.201 ms for the PISD1, PISD2 and PISD3 cases, respectively, than for the normal one. The results show that aqueous layers within the myelin sheath do not modulate the normal strength-duration time constant and rheobasic current, which are 0.291 ms and 0.388 nA, respectively.

For the ISD1, ISD2 and ISD3 cases, rheobasic currents increase in both cases without vs. with aqueous layers within the myelin sheath and they are 0.568/0.605, 0.709/0.805 and 1.444/1.718 nA, respectively. However, the rheobasic currents are increased versus decreased 0.842/0.793, 1.149/1.084 and 1.741/1.654 nA for the PSD1, PSD2 and PSD3 cases without vs. with aqueous layers, respectively. They significantly increase in the PISD cases with the increase of the degree of demyelination, without vs. with aqueous layers. Rheobasic currents are 1.020, 1.487 and 1.616 nA for the PISD1, PISD2 and PISD3 cases, respectively.

Results show that aqueous layers within the internodally and paranodally systematically demyelinated sheaths modulate the strength-duration properties (time constants and rheobases) of human motor nerve fibres. The main excitability parameter, differentiating the ISDs from the PSDs, is the strength-duration time constant. It is longer for the ISDs and shorter for the PSDs, than the normal one. According to this parameter, the PSDs are regarded by us as specific indicators for chronic demyelinating neuropathies, whereas, the ISDs are regarded by us as specific indicators for hereditary demyelinating neuropathies, as was already proven and discussed in Chapter III. The addition of aqueous layers improves the strength-duration time constants for both demyelinated types. However, they are still longer in the ISD cases than the normal one. Such strength-duration time constants are more characteristic for patients with CMT1A (Nodera et al. 2004, Nodera and Kaji 2006). The shorter strength-duration time constants in the PSD cases are characteristic for patients with CIDP (Cappelen-Smith et al. 2001, Nodera and Kaji 2006). Similarly, the strength-duration time constant histograms in the ISDs and PSDs are changed in totally opposite directions, because of the different electrogenesis of different demyelinations (Fig. 37). The strength-duration time constant is based on the passive (RC) membrane parameters in parallel, according to the classical theory, and it is calculated in our studies by a transfer standard parabola, which fits the charge-duration curves (Fig. 36). The slopes of the charge-duration curves in the PSD and in the ISD cases are in opposite directions, because of the differently changed RC parameters. According to the classical theory, there is an inverse relationship between the strength-duration time constant and rheobasic current, and the results presented here show that this is also valid when the myelin aqueous layers are taken into account.

The results presented here also show that aqueous layers within the myelin sheath do not modulate axonal excitability properties such as the strength-duration time constants and rheobasic currents of human motor nerve fibres when the systematic demyelinations are mixed, i.e., paranodal internodal (PISDs). The reciprocally opposed effects of the aqueous layers within the myelin sheath on the strength-duration time constants and rheobasic currents in the ISD and PSD cases are neutralized when the demyelinations are heterogeneous. Moreover, the PISD fibres without vs. with aqueous layers within the myelin sheath behave like PSD fibres, since the myelin reduction slightly increases the effect of the paranodal demyelination on these axonal excitability properties. The effect of the reduced paranodal seal resistance, additionally increased by reduced myelin sheath without vs. with aqueous layers could be consistent with the pathophysiology in patients with CIDP subgroups (Sung et al. 2004) which have heterogeneous demyelinations.

The excitability changes during 100 ms recovery cycles in the normal, ISD (a), PSD (b) and PISD (c) cases without (*dotted lines*) and with (*continuous lines*) aqueous layers within the myelin sheath are illustrated in Fig. 38 for human motor nerve fibres. The results show significant differences in the axonal excitability changes between the normal and demyelinated cases. Recovery cycles are the same in the normal case without and with aqueous layers. Normal axons are initially unexcitable, subsequently are excitable with a raised threshold, and after about 3 ms they are superexcitable. Results also show that the axonal excitability changes in the ISD cases are similar. The axonal types are also initially unexcitable, then superexcitable. The superexcitability is usually followed by late subexcitability of axons. The refractoriness (the increase in threshold current during the relative refractory period) in these demyelinated cases with aqueous layers is slightly higher or near normal. Maximal values of the refractoriness are 14% for the cases ISD1 and 16% for ISD2 without aqueous layers (not illustrated in Fig. 38a). The relative refractory periods are similar and last less than 3 ms. Axons without vs. with aqueous layers within the myelin sheaths have substantially greater superexcitability and less late subexcitability than those of the normal axon. However, there is an additional increase of axonal superexcitability in the demyelinated cases when the myelin aqueous layers are taken into account.

Compared to the normal case, the recovery cycles in the investigated PSD cases are mainly without subexcitable relative refractory periods. The subexcitable relative refractory periods are characteristic for considerably mild (20–30%) demyelinations (Stephanova and Daskalova 2005a). Demyelinated axons without aqueous layers have substantially greater superexcitability and less late subexcitability than those of the normal axon. There is an additional slight decrease of axonal superexcitability in the PSD1 case with aqueous layers. However, the effect of the aqueous layers on the axonal superexcitability decreases with the increase of the demyelination from PSD2 to PSD3 and could be considered as negligible. An overlap of the recovery cycles is obtained for the normal and PISD cases without and with aqueous layers.

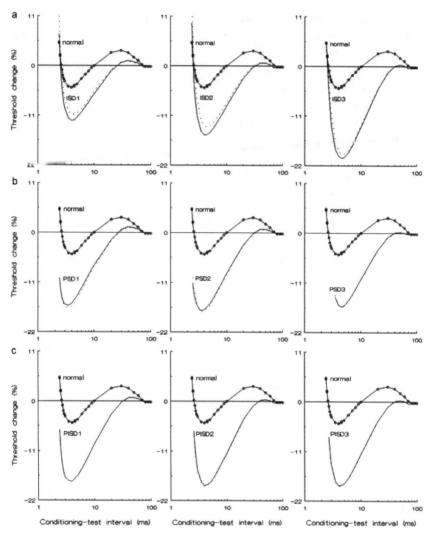

Fig. 38. Comparison of recovery cycles of human motor nerve fibres in the normal, ISD **(a)**, PSD **(b)** and PISD **(c)** cases without (*dotted lines*) and with (*continuous lines*) aqueous layers within the myelin sheath. For all investigated cases, the y-axis is defined as $100 \times (I_{test} - I_{cond})/I_{cond}$ (%), where I_{test} (nA) and I_{cond} (nA) are the threshold currents of the testing and conditioning pulses, respectively.

Abnormally greater superexcitabilities are found for the PISD1, PISD2 and PISD3 types. In these demyelinated cases compared with the normal case, the recovery cycles are without relative refractory periods and only have superexcitable periods. The refractoriness in the ISD cases with aqueous layers is slightly higher or near normal. Such refractoriness is more characteristic for patients with CMT1A (Nodera et al. 2004). The greater superexcitability of the PSD axons is characteristic for patients with CIDP. Changes in the excitability properties obtained in the PISD cases are in good accordance with the data from patients with CIDP subgroups (Sung et al. 2004).

In summary, the results presented here show that aqueous layers, when are taken into account, have an additional effect on the simulated cases. All excitability properties, except for the refractoriness and strength-duration time constants worsen in the simulated hereditary demyelinating neuropathies, when the myelin lamellae and their corresponding aqueous layers are reduced. In these simulated demyelinated neuropathies, excitability property changes are bigger for the accommodative and adaptive processes than for the propagating processes. The mild internodal systematic demyelinations studied here are characteristic for the CMT Type1 such as CMT1A, whereas the severe internodal systematic demyelinations studied here are more characteristic for the CMT Type3 such as Dejerine-Sottas syndrome (DSS).

In the simulated chronic inflammatory demyelinating polyneuropathies such as CIDP, the addition of aqueous layers enhances—and can even restore—action potential propagation in some severely paranodally demyelinated axons. Aqueous layers improve all excitability properties in the simulated chronic inflammatory demyelinating polyneuropathies. Excitability property changes are bigger for the propagating processes than for the accommodative and adaptive processes in these simulated cases.

Aqueous layers do not modulate the excitability membrane properties in the simulated CIDP subtypes. The reciprocally opposed effects of the aqueous layers on these properties are neutralized, when the demyelinations are heterogeneous.

Results confirm that the transition from conduction slowing to conduction block leads to amplification of the degree of excitability property changes, as the direction of these changes is maintained. This is valid for both cases without and with aqueous layers. However, if there is complete conduction block, the threshold tracking technique is practically impossible to use. Consequently, the results presented here not only widen the boundaries of the objective study of mechanisms in hereditary and chronic demyelinating neuropathies, but also have a predictive value.

Results also show that when the reduction value is under 70% the effect of the myelin sheath aqueous layers on the excitability membrane properties is negligible for all investigated subtypes. We suggest that the effect of the aqueous layers on the excitability properties of the simulated GBS and MMN will also be negligible, as the excitability property changes in the focally demyelinated subtypes are vastly smaller than those in the systematically demyelinated subtypes. However, even differences of 1–3%, as a result of the effect of nanostructures (such as myelin sheath aqueous layers) on the excitability membrane properties investigated here could be regarded as significant. The results presented in Chapter IV are based on our previous studies (Krustev et al. 2010, 2012, Stephanova et al. 2011a,b).

References

Alaedini, A. and H.W. Sander, A.P. Hays, and N. Latov. 2003. Anti-ganglioside antibodies in multifocal acquired sensory and motor neuropathy. Arch. Neurol. 60: 42–46.

Alexandrov, A.S. and L.C. Christova, B.A. Ishpekova, and D.S. Kosarov. 2003. A study of A waves in patients with facial nerve neuritis. Comptes rendus de l'Academie bulgare des Sciences 56(11): 97–100.

Arasaki, K. and M. Tamaki. 1998. A loss of functional spinal alpha motor neuronal in amyotrophic lateral sclerosis. Neurology 51: 603–605.

Arbuthnott, E.R. and I.A. Boyd, and K.U. Kalu. 1980. Ultrastructural dimensions of myelinated peripheral nerve fibres in the cat and their relation to conduction velocity. J. Physiol. 308: 125–157.

Awiszus, F. 1990. Effects of paranodal potassium permeability on repetitive activity of mammalian nerve fiber models. Biol. Cybern. 64: 69–76.

Baker, M. and H. Bostock, P. Grafe, and P. Martius. 1987. Function and distribution of three types of rectifying channel in rat spinal root myelinated axons. J. Physiol. (Lond.) 383: 45–67.

Baraitser, M. The genetics of neurological disorders. pp. 238–239. In: A. Motulski, M. Bobrow, P. Harper, and C. Scriver [eds.]. 1990. Oxford monographs on Medical genetics, Oxford University Press, Oxford.

Barnett, M.H. and J.D. Pollard, L. Davies, and J.G. McLeod. 1998. Cyclosporin A in resistant chronic inflammatory demyelinating polyradiculoneuropathy. Muscle Nerve 21: 454–460.

Barohn, R.J. and J.T. Kissel, J.R. Warmolts, and J.R. Mendell. 1989. Chronic inflammatory demyelinating polyradiculoneuropathy: clinical characteristics, course and recommendations for diagnostic criteria. Arch. Neurol. 46: 878–884.

Barrett, E.F. and J.N. Barrett. 1982. Intracellular recordings from vertebrate myelinated axons: mechanism of the depolarizing after-potentials. J. Physiol. (Lond.) 323: 117–144.

Bensimon, G. and L. Lacomblez, and V. Meininger. 1994. A controlled trial of riluzole in amyotrophic lateral sclerosis. ALS/riluzole study group. N. Engl. J. Med. 330: 585–591.

Benstead, T.J. and N.L. Kuntz, R.G. Miller, and J.R. Daube. 1990. The electrophysiologic profile of Dejerine-Sottas disease (HMSN III). Muscle Nerve 13(7): 586–592.

Berger, P. and A. Niemann, and U. Suter. 2006. Schwann cells and the pathogenesis of inherited motor and sensory neuropathies (Charcot-Marie-Tooth disease). Glia 54: 243–257.

Bertam, R. and M.J. Butte, T. Keimel, and A. Sherman. 1995. Topological and phenomenological classification of bursting oscillators. Bull. Math. Biol. 57: 413–449.

Berthold, C.H. Morphology of normal peripheral axons. pp. 3–63. *In:* S. Waxman [ed.] 1978. Physiology and pathology of axons. Raven Press, New York, USA.

Berthold, C.H. and T. Carlstedt, and O. Corneliuson. The central-peripheral transition zone. pp. 73–80 *In.* P.J. Dyck, and P.K. Thomas [eds.]. 1983. Peripheral Neurophaty. WB Saunders Company, Philadelphia.

Berthold, C.H. and M. Rydmark. 1983. Anatomy of the paranode-node-paranode region in the cat. Experientia 39: 964–976.

Bhat, M.A. 2003. Molecular organization of axo-glial junctions. Curr. Opin. Neurobiol. 13: 552–559.

Bhat, M.A. and J.C. Rios, Y Lu, Garcia-Fresco, W. Ching, M.S. Martin, J. Li, S. Einheber, M. Chesler, J. Rosenbluth, J.L. Salzer, and H.J. Bellen. 2001. Axon-glia interactions and the domain organization of myelinated axons require NeurexinIV/Caspr/Paranodin. Neuron 30: 369–383.

Birchmeier, C. and K.A. Nave. 2008. Neuregulin-1, a key axonal signal that drives Schwann cell growth and differentiation. Glia 56(14): 1491–1497.

Birouk, N. and R. Gouider, E. Le Guern, M. Gugenheim, S. Tardieu, T. Maisonobe, N. Le Forestier, Y. Agid, A. Brice, and P. Boushe. 1997. Charcot-Marie-Tooth disease type 1A with 17p11.2 duplication. Clinical and electrophysiological phenotype study and factors influencing disease severity in 119 cases. Brain 120: 813–823.

Blaurock, A.E. 1971. Structure of nerve myelin membrane: proof of the low resolution profile. J. Membr. Biol. 56: 35–52.

Blight, A. 1985. Computer simulation of action potentials and after-potentials in mammalian myelinated axons: the case for a lower resistance myelin sheath. Neuroscience 15: 13–31.

Blight, A. and S. Someya. 1985. Depolarizing afterpotential in myelinated axons of mammalian spinal cord. Neuroscience 15: 1–12.

Boërio, D. and A. Créange, J.Y. Hogler, A. Guéguen, D. Berland, and J.P. Lefaucheur. 2010. Nerve excitability changes after intravenous immunoglobulin infusions in multifocal motor neuropathy and chronic inflammatory demyelinating neuropathy. J. Neurol. Sci. 292: 63–71.

Bostock, H. Mechanisms of accommodation and adaptation in myelinated axons. pp. 311–327. *In:* S.G. Waxman, J.D. Kocsis, and P.K. Stys [eds.]. 1995. The axon. Oxford University Press, New York, USA.

Bostock, H. and M. Baker. 1988. Evidence for two types of potassium channel in human motor axons *in vivo*. Brain Res. 462: 354–358.

Bostock, H. and M. Baker, and G. Reid. 1991. Changes in excitability of human motor axons underlying post-ischaemic fasciculation: evidence for two stable states. J. Physiol. 441: 537–557.

Bostock, H. and D. Burke, and J.P. Hales. 1994. Differences in behavior of sensory and motor axons following release of ischaemia. Brain 117: 225–234.

Bostock, H. and K. Cikurel, and D. Burke. 1998. Threshold tracking techniques in the study the human peripheral nerve. Muscle Nerve 21: 137–158.

Bostock, H. and J.C. Rothwell. 1997. Latent addition in motor and sensory fibres of human peripheral nerve. J. Physiol. (Lond.) 498: 277–294.

Bostock, H. and M.K. Sharief, G. Reid, and N.M.F. Murray. 1995. Axonal ion channel dysfunction in amyotrophic lateral sclerosis. Brain 118: 217–225.

Brismar, T. and J.R. Schwarz. 1985. Potassium permeability in rat myelinated nerve fibres. Acta Physiol. Scand. 124: 141–148.

Brown, R.H. 1997. Amyotrophic lateral sclerosis: Insights from genetics. Arch. Neurol. 54: 1246–1250.

Butt, A.M. and A. Duncan, M.F. Hornby, S.L. Kirvell, A. Hunter, J.M. Levine, and M. Barry. 1999. Cells expressing the NG2 antigen contact nodes of Ranvier in adult CNS white matter. Glia 26(1): 84–91.

Cappelen-Smith, C. and S. Kuwabara, CS-Y. Lin, I. Mogyoros, and D. Burke. 2001. Membrane properties in chronic inflammatory demyelinating polyneuropathy. Brain 124: 2439–2447.

Carvalho, A.A and A. Vital, X. Ferrer, P. Latour, A. Lagueny, C. Brechenmacher, and C. Vital. 2005. Charcot-Marie-Tooth disease type 1A: clinicopathological correlations in 24 patients. J. Peripher. Nerv. Syst. 10: 85–92.

Chang, A.P. and J.D. England, C.A. Garcia, and A.J. Summer. 1998. Focal conduction block in n-hexane polyneuropathy. Muscle Nerve 21: 964–969.

Chaudhry, V. and A.M. Corse, D.R. Cornblath, R.W. Kuncl, D.B. Drachman, M.L. Freimer, R.G. Miller, and J.W. Griffin. 1993. Multifocal motor neuropathy: Response to human immune globulin. Ann. Neurol. 33: 237–242.

Ching, W. and G. Zanazzi, S.R. Levinson, and J.L. Salzer. 1999. Clustering of neuronal sodium channels requires contact with myelinating Schwann cells. J. Neurocytol. 28: 295–301.

Chiu, S.Y. and J.M. Ritchie. 1981. Evidence for presence of potassium channels in the paranodal region of acutely demyelinated mammalian single nerve fibres. J. Physiol. 313: 415–437.

Chiu, S.Y. and J.M. Ritchie. 1984. On the physiological role of internodal potassium channels and the security of conduction in myelinated nerve fibres. Proc. R. Soc. Lond. B. 220: 415–422.

Chiu, S.Y. and W. Schwarz. 1987. Sodium and potassium currents in acutely demyelinated internodes of rabbit sciatic nerve. J. Physiol. 391: 631–649.

Chiu, S.Y. and L. Zhou, C.L. Zhang, and A. Messing. 1999. Analysis of potassium channel function in mammalian axons by gene knockings. J. Neurocytol. 28: 349–364.

Chowdhury, D. and A. Arora. 2001. Axonal Guillain-Barré syndrome: a critical review. Acta Neurol. Scan. 103: 267–277.

Christova, L.G. and A.S. Alexandrov, B.A. Ishpekova, and D.S. Kosarov. 2001. Comparative analyse of single motor unit pattern in healthy subjects and patients with neuromuscular disorders. Acta Physiol. Pharmacol. Bulg. 26: 59–62.

Clark, J.W. and R. Plonsey. 1968. The extracellular potential field of the single active nerve fiber in a volume conductor. Biophys. J. 8: 842–864.

Clark, J.W. and R. Plonsey. 1970. A mathematical study of nerve fibre interaction. Biophys. J. 10: 973–957.

Cleveland, D.W. 1999. From Charcot to SOD1: mechanisms of selective motor neuron death in ALS. Neuron 24: 515–520.

Coggesall, R.E. and J.D. Coulter, and W.D. Willis. 1974. Unmyelinated axons in the ventral roots of the cat lumbo-sacral enlargement. J. Comp. Neurol. 153: 39–58.

Coleman, M.P. and M.F. Freeman. 2010. Wallerian degeneration WldS and Nmnat, Annual Review of Neuroscience 33: 245–267.

Conradi, S. and L. Grimby, and G. Lundemo. 1982. Pathophysiology of fasciculations in ALS as studied by electromyography of single motor units. Muscle Nerve 5: 202–208.

Court, F.A. and W.T. Hendriks, H.D. MacGillavry, J. Alvarez, and J. van Minnen. 2008. Schwann cell to axon transfer of ribosomes: toward a novel understanding of the role of glia in the nervous system. J. Neurosci. 28: 11024–11029.

Daskalova, M. and D.I Stephanova. 2001. Strength-duration properties of human myelinated motor and sensory axons in normal case and in amyotrophic lateral Sclerosis. Acta Physiol. Pharmacol. Bulg. 26: 1–4.

Daube, J.R. 2000. Electrodiagnostic studies in amyotrophic lateral sclerosis and other motor neuron disorders. Muscle Nerve 23: 1488–1502.

Dimitrov, A.G. 2000. The effect of a near-membrane volume on generation of action potentials in myelinated nerve fibres. Neurophysiology 32: 228–228.

Dimitrov, A.G. 2005. Internodal sodium channels ensure active processes under myelin manifesting in depolarizing afterpotentials. J. Theor. Biol. 234: 451–462.

Dimitrov, A.G. 2009. A possible mechanism of repetitive firing of myelinated axon. Pflügers Arch. Eur. J. Physiol. 458: 547–561.

Dioszeghy, P. and E. Stålberg. 1992. Changes in motor and sensory nerve conduction parameters with temperature in normal and diseased nerve. Electroencephalogr. Clin. Neurophysiol. 85: 229–235.

Dittrich, F. and G. Ochs, A. Grosse-Wilde, U. Berweiler, Q. Yan, J.A. Miller, K.V. Toyka, and M. Sendtner. 1996. Pharmacokinetics of intrathecally applied BDNF and effects on spinal motoneurons. Exp. Neurol. 141(2): 225–239.

Dodge, F. and B. Frankenhaeuser. 1959. Sodium currents in the myelinated nerve fiber of Xenopus laevis investigated with the voltage clamp technique. J. Physiol. 148: 188–200.

Donofrio, P.D. and J.W. Albers. 1990. Polyneuropathy: classification by nerve conduction studies and electromyography. Muscle Nerve 13: 889–903.

Douglas, D.S. and B. Popko. 2009. Mouse forward genetics in the study of the peripheral nervous system and human peripheral neuropathy. Neurochem. Res. 34(1): 124–137.

Dučić, T. and S. Quintes, K.A. Nave, J. Susini, M. Rak, R. Tucoulou, M. Alevra, P. Guttman, and T. Salditt. 2011. Structure and composition of myelinated axons: A multimodal synchrotron spectro-microscopy study. Journal of Structural Biology 173: 202–212.

Dupree, J.L. and J.A. Girault, and B. Popko. 1999. Axo-glial interactions regulate the localization of axonal paranodal proteins. J. Cell Biology 147(6): 1145–1151.

Dyck, P.J. and P. Chance, R. Lebo, and A.J. Camey. Hereditary motor and sensory neuropathies. pp. 1094–1136. *In:* P.J. Dyck, P.K. Thomas, J.W. Griffin,

P.A. Low, and J.F. Poduslo [eds.]. 1993. Peripherial neuropathy, W.B. Saunders, Philadelphia.

Dyck, P.J. and E.H. Lambert. 1968a. Lower motor and primary sensory neuron diseases with peroneal muscular atrophy. Neurologic, genetic, and electrophysiologic findings in hereditary polyneuropathies. Arch. Neurol. 18(6): 603–618.

Dyck, P.J. and E.H. Lambert. 1968b. Lower motor and primary sensory neuron diseases with peroneal muscular atrophy. Neurologic, genetic, and electrophysiologic findings in various neuronal degenerations. Arch. Neurol. 18(6): 619–625.

Dyck, P.J. and E.H. Lambert. 1974. Compound nerve action potentials and morphometry. Electroencephalogr. Cin. Neurophysiol. 36: 573–574.

Dyck, P.J. and W.J. Litchy, K.M. Kratz, G.A. Suarez, P.A. Low, A.A. Pineda, A.J. Eindeback, J.L. Karens, and P.C. O'Brien. 1994. A plasma exchange versus immune globulin infusion trail in chronic inflammatory demyelinating polyradiculoneuropathy. Ann. Neurol. 36: 838–845.

Dyck P.J. and P.C. O'Brien, K.F. Oviatt, R.P. Dinapoll, J.R. Daube, J.D. Bartleson, B. Mokri, T. Swift, P.A. Low, and A.J. Windebank. 1982. Prednisone improves chronic inflammatory demyelinating neuropathy more than no treatment. Ann. Neurol. 10: 136–141.

Dzhashiashvili, Y. and Y. Zhang, J. Galinska, I. Lam, M. Grumet, and J.L. Salzer. 2007. Nodes of Ranvier and axon initial segments are ankyrin G-dependent domains that assemble by distinct mechanisms. J. Cell Biol. 177(5): 857–870.

Eftimov, F. and J.B. Winer, M. Vermeulen, R. de Haan, and I.N. van Schaik. 2009. Intravenous immunoglobulin for chronic inflammatory demyelinating polyradiculoneuropathy. Cochrance Database Syst. Rev. 21(1): CD001797.

Fabrizi, G.M. and T. Cavallaro, C. Angiari, I. Cabrini, F. Taioli, G. Malerba, L. Bertolasi, and N. Rizzuto. 2007. Charcot-Marie-Tooth disease type 2E, a disorder of the cytoskeleton, Brain 130: 394–403.

Feasby, T.E. Treatment of chronic inflammatory demyelinating neuropathy (CIDP). pp. 884–886. In: J. Kimura and H. Shibasaki [eds.]. 1996. Recent advance in clinical Neurophysiology. Elsevier Science BV, Amsterdam.

Feasby, T.E. and J.J. Gilbert, W.F. Brown, C.F. Bolton, A.F. Hahn, W.F. Koopman, and D.W. Zochodne. 1986. An acute axonal form of Guillain-Barré polyneuropathy. Brain 109: 1115–1126.

Feinberg, D.M. and D.C. Preston, J.M. Shefner, and E.L. Logigian 1999. Amplitude-dependent slowing of conduction in amyotrophic lateral sclerosis and polyneuropathy. Muscle Nerve 22: 937–940.

Finsterer, J. and A. Fuglsang-Frederiksen, and B. Mamoli. 1997. Needle EMG of the tongue: motor unit action potential versus peak ratio analysis in limb and bulbar onset amyotrophic lateral sclerosis. J. Neurol. Neurosurg. Psychiatry 63: 175–180.

FitzHugh, R. 1962. Computation of impulse initiation and salutatory conduction in a myelinated nerve fiber. Biophys. J. 2: 11–21.

Frankenhaeuser, B. and A.F. Huxley. 1964. The action potential in the myelinated nerve fiber of Xenopus Laevis as computed on the basis of voltage clamp data. J. Physiol. (Lond.) 171: 302–315.

Frankenhaeuser, B. and B. Waltman. 1959. Membrane resistance and conduction velocity of large myelinated nerve fibres from Xenopus laevis. J. Physiol. 148: 677–682.

Friede, R.L. and W. Beuche. 1985. A new approach toward analyzing peripheral nerve fiber populations. I Variance in sheath thickness corresponds to different properties of the internode. J. Neurophatol. Exp. Neurol. 44: 60–72.

Friede, R.L. and R. Bischhauser. 1982. How are sheath dimensions affected by axon clibre and internodal length? Brain Res. 235: 335–350.

Friede, R.L. and T. Meier, and M. Diem. 1981. How is the exact length of an internode determined? J. Neurol. Sci. 50: 217–228.

Gabreels-Festen, A. 2002. Dejerine-Sottas syndrome grown to maturity: overview of genetic and morphological heterogeneity and follow-up of 25 patients, J. Anat. 200(4): 341–356.

Ganapathy, L. and J.W. Clark. 1987. Extracellular currents and potentials of the active myelinated nerve fibre. Biophys. J. 52: 749–761.

Garbay, B. and A.M. Heape, F. Sargueil, and C. Cassagne. 2000. Myelin synthesis in the peripheral nervous system. Prog. Neurobiol. 61: 267–304.

Gilliatt, R.W. and P.K. Thomas. 1957. Extreme slowing of nerve conduction in peroneal muscular atrophy. Ann. Phys. Med. 4(3): 104–106.

Goldfinger, M.D. 2000. Computation of High Safety Factor Impulse Propagation at Axonal Branch Points. Neuro Report 11(3): 449–456.

Goldfinger, M.D. 2005a. Highly-efficient propagation of random impulse trains across unmyelinated axonal branch points: modifications by periaxonal K$^+$ accumulation and sodium channel kinetics. pp. 479–529. In: G.N. Reeke, R.R. Poznanski, K.A. Lindsay, J.R. Rosenberg, and O. Sporns. [eds.]. 2005a. Modeling in the Neurosciences. Taylor and Francis, Boca Raton, London, New York, Singapore.

Goldfinger, M.D. 2005b. Rallian "Equivalent" cylinders reconsidered: Comparisons with literal compartments. J. Integr. Neurosci. 4(2): 227–263.

Goldfinger, M.D. 2009. Probability distributions of Markovian Na channel states during propagating axonal impulses with or without recovery supernormality. J. Integr. Neurosci. 8: 203–221.

Goldman, L. and J.S. Albus. 1968. Computation of impulse conduction in myelinated fibers: theoretical basis of velocity-diameter relation. Biophys. J. 8: 596–607.

Gong, Y. and Y. Tagawa, M.P.T. Lunn, W. Laroy, M. Haffer-Lauc, C.Y. Li, J.W. Griffin, R.L. Schnaar, and K.A. Sheikh. 2002. Localization of major gangliosides in the PNS: implications for immune neuropathies. Brain 125(11): 2491–25016.

Gorson, K.C. and G. Allam, and A.H. Ropper. 1997. Chronic inflammatory demyelinating polyneuropathy: Clinical features and response to treatment in 67 consecutive patients with and without a monoclonal gammopathy. Neurology 48: 321–327.

Gorson, K.C. and A.H. Ropper, L.S. Adelman, and D.H. Weinberg. 2000. Influence of diabetes mellitus on chronic inflammatory demyelinating polyneuropathy. Muscle Nerve 23: 37–48.

Gorson, K.C. and A.H. Ropper, B.D. Clark, R.B. Dew, D. Simovic, and G. Allam. 1998. Treatment of chronic inflammatory demyelinating polyneuropathy with interferon-α 2a. Neurology 50: 84–87.

Gow, A. and J. Devaux. 2008. A model of tight junction function in central nervous system myelinated axons. Neurol Glia Biology, Department of Biomedical Engineering, Case Western Reserve University 4(4): 307–317.

Guckenheimer, J. and R. Harris, J. Peck, and A. Willms. 1997. Bifurcation, bursting and spike frequency adaptation. J. Comput. Neurosci. 4: 257–277.

Guertin, A.D. and D.P. Zhang, K.S. Mak, J.A. Alberta, and H.A. Kim. 2005. Microanatomy of axon/glial signaling during Wallerian degeneration. Journal of Neuroscience 25: 3478–3487.

Guiloff, R.J. and H. Modarres-Sadeghi. 1992. Voluntary activation and fiber density of fasciculations in motor neuron disease. Ann. Neurol. 31: 416–424.

Halter, J. and J. Clark. 1991. A distributed-parameter model of the myelinated nerve fiber. J. Theor. Biol. 148: 345–382.

Harding, A.E. and P.K. Thomas. 1980. Genetic aspects of hereditary motor and sensory neuropathy (types I and II). J. Med. Genet. 17(5): 329–336.

Hartung, H.P. and J.D. Pollar, G.K. Harvey, and K.V. Toyka. 1995. Immuno-pathogenesis and treatment of the Guillain-Barré syndrome—Part I. Mescle Nerve 18: 137–153.

Hattori, N. and M. Yamamoto, T. Yoshihara, H. Koike, M. Nakagawa, H. Yoshikawa, A. Ohnishi, K. Hayasaka, O. Onodera, M. Baba, H. Yasuda, T. Saito, K. Nakashima, J. Kira, R. Kaji, N. Oka, and G. Sobue. 2003. Demyelinating and axonal features of Charcot-Marie-Tooth disease with mutations of myelin-related proteins (PMP22, MPZ and Cx32): a clinicopathological study of 205 Japanese patients, Brain 126: 134–151.

Henriksen, J. 1956. Conduction velocity of motor nerves in normal subjects and patients with neuromuscular disorders. Ph. Thesis, University of Minesota, Minesota.

Heredia, A. and C.C. Bui, U.S. Suter, P. Young, and T.E. Schäffer. 2007. AFM combines functional and morphological analysis of peripheral myelinated and demyelinated nerve fibers. Neuro Image 37: 1218–1226.

Hill, A.V. 1936. Excitation and accommodation in nerve. Proc. R. Soc. B 119: 305–355.

Hirose, G. and J. Kimura, K. Arimura, M. Baba, H.G. Hara, K. Hirayama, R. Kaji, M. Kanda, T. Kobayashi, H. Kowa, Y. Kuroiwa, T. Mezaki, G. Sobue, and N. Yanagisawa. A trial study of chronic inflammatory demyelinating polyneuropathy with intravenous immunoglobulin. pp. 879–883. In: J. Kimura and H. Shibasaki [eds.]. 1996. Recent advance in clinical neurophysiology. Elsevier Science BV, Amsterdam.

Hodgkin, A. and A. Huxley. 1952a. Currents carried by sodium and potassium ions through the membrane of the giant axon of Loligo. J. Physiol. 116: 449–472.

Hodgkin, A. and A. Huxley. 1952b. The components of membrane conductance in the giant axon of Loligo. J. Physiol. 116: 473–496.

Hodgkin, A. and A. Huxley. 1952c. The dual effect of membrane potential of sodium conductance in the giant axon of Loligo. J. Physiol. 116: 497–499.

Hodgkin, A. and A. Huxley. 1952d. A quantitative description of membrane current and its application to conductance and excitation in nerve. J. Physiol. 117: 500–544.

Hodgkin, A. and A. Huxley, and B. Katz. 1949. Ionic currents underlying activity in giant axon of the squid. Arch. Sci. Physiol. 3: 129–150.

Hodgkin, A. and A. Huxley, and B. Katz. 1952. Measurement of current-voltage relations in the membrane of the giant axon of Loligo. J. Physiol. 116: 424–448.

Hodgkinson, S.J. and J.D. Pollard and J.G. McLeod. 1990. Cyclosporin A in the treatment of chronic inflammatory demyelinating polyradiculo-neuropathy. J. Neurol. Neurosurg. Psychiatry 53: 327–330.

Houlden, H. and M.M. Reilly. 2006. Molecular genetics of autosomal-dominant demyelinating Charcot-Marie-Tooth disease. Neuromole-cular Med. 8: 43–62.

Hughes, R.A. 2002. Systematic reviews of treatment of inflammatory demyelinating neuropathy. J. Anat. 200(4): 331–339.

Inouye, H. and D.A. Kirschner. 1988a. Membrane interaction in nerve myelin. 1. Determination of surface-charge from effects of pH and ionic strength on period. Biophys. J. 53: 235–245.

Inouye, H. and D.A. Kirschner. 1988b. Membrane interaction in nerve myelin. 1. Determination of surface-charge from biochemical data. Biophys. J. 53: 247–260.

Iwasaki, Y. and T. Shiojima, M. Kinoshita, and K. Ikeda. 1998. SR57746A: asurvival factor for motor neurons *in vivo*. J. Neurol Sci. 160 (suppl. 1): S92–S96.

Jessen, K.R. and R. Mirsky. 2005. The origin and development of glial cells in peripheral nervous system. Nat. Rev. Neurosci. 6: 671–682.

Johannsen, G. 1986. Line source models of active fibres. Biol. Cybern. 54: 151–158.

Kaji, R. 2003. Physiology of conduction block in multifocal motor neuropathy and other demyelinating neuropathies. Muscle Nerve 27: 285–296.

Kaji, R. and N. Hirota, N. Oka, N. Kohara, T. Watanabe, T. Nishio, and J. Kimura. 1994. Anti-GM1 antibodies and impaired blood-nerve barrier may interfere with remyelination in multifocal motor neuropathy. Muscle Nerve 17: 108–110.

Kaji, R. and H. Shibasaki, and J. Kimura. 1992. Multifocal demyelinating motor neuropathy: Cranial nerve involvement and immunoglobulin therapy. Neurology 42: 506–509.

Kaplan, M.R. and A. Meyer-Franke, S. Lambert, V. Bennett, I.D. Duncan, S.R. Levinson, and B.A. Barres. 1997. Induction of sodium channel clustering by oligodendrocytes. Nature 386: 724–728.

Katz, J.S. and D.S. Saperstain, G. Gronseth, A.A. Amato, and R.J. Barohn. 2000. Distal acquired demyelinating symmetric neuropathy. Neurology 54: 615–620.

Katz, J.S. and G.I. Wolfe, W.W. Bryan, C.E. Jackson, A.A. Amato, and R. Barohn. 1997. Electrophysiologic finings in multifocal motor neuropathy. Neurology 48: 700–707.

Kearney, J.A. and D.A. Buchner, G. De Haan, M. Adamska, M. Levin, A.R. Furay, R.L. Albin, J.M. Jones, M. Montal, and M.J. Stevens. 2002. Molecular and pathological effects of a modifier gene on deficiency of the sodium channel Scn8a (Na(v)1.6). Hum. Mol. Genet. 11: 2765–2775.

Kiernan, M.C. and D. Burke, K.V. Andersen, and H. Bostock. 2000. Multiple measures of axonal excitability: a new approach in clinical testing. Muscle Nerve 23: 399–409.

Kiernan, M.C. and K. Cikurel, and H. Bostock. 2001. Effects of temperature on the excitability properties of human motor axons. Brain 124: 816–825.

Kiernan, M.C. and J.M. Guglielmi, R. Kaji, N.M.F. Murray, and H. Bostock. 2002. Evidence for axonal membrane hyperpolarization in multifocal motor neuropathy with conduction block. Brain 125: 664–675.

Kimura, J. Multifocal motor neuropathy and conduction block. pp. 57–72. *In*: J. Kimura and R. Kaji. [eds.]. 1997. Physiology of ALS and related disorders. Elsevier Science BV, Amsterdam.

Kimura, J. Inflammatory, infective and autoimmune neuropathies. pp. 661–710. *In*: J. Kimura [ed.]. 2001. Electrodiagnosis in diseases of nerve and muscle. Principles and Practice, University Press, Oxford.

Kirschner, D.A. and D.L. Caspar. 1972. Comparative diffraction studies on myelin membranes. Ann. N .Y. Acad. Sci. 195: 309–317.

Kirschner, D.A. and A.L. Ganser, and D.L. Caspar. Diffraction studies of molecular organization and membrane interactions in myelin. pp. 51–95. *In*: P. Morrell [ed.]. 1984. Myelin, Plenum Press, New York.

Kocsis, J.D. and S.G. Waxman, C. Hildebrand, and J.A. Ruiz. 1982. Regenerating mammalian nerve fibers: changes in action potential waveform and firing characteristics following blockage of potassium conductance. Proc. R. Soc. Lond. B Biol. Sci. 217: 77–87.

Koles, Z.J. and M. Rasminsky. 1972. A computer simulation of conduction in demyelinated nerve fibres. J. Physiol. (Lond.) 227: 351–364.

Köller, H. and B.C. Kieseier, S. Jander, and H.P. Hartung. 2005. Chronic inflammatory demyelinating polyneuropathy, Review Article. N. Eng. J. Med. 352: 1343–1356.

Krarup, C. and J.D. Stewart, A.J. Sumner, A. Pestronk, and S.A. Lipton. 1990. A syndrome of asymmetric limb weakness with motor conduction block. Neurology 40: 118–127.

Krustev, S. and M. Daskalova, and D. Stephanova. 2010. Myelin sheath aqueous layers do not modulate membrane fibre properties of simulated cases of paranodal internodal systematic demyelinations. Compt. rend. Acad. bulg. Sci. 63(12): 1845–1852.

Krustev, S.M. and N. Negrev, and D.I. Stephanova. 2012. The strength-duration properties in simulated demyelinating neuropathies depend on the myelin sheath aqueous layers. Scripta Scientifica Medica 44(1): 117–123.

Kuwabara, S. and H. Bostock, K. Ogawara, J.Y. Sung, K. Kanai, M. Mori, T. Hattori, and D. Burke. 2003. The refractory period of transmission is impaired in axonal Guillain-Barré syndrome. Muscle Nerve 28: 683–689.

Kuwabara, S. and M. Nakajima, Y. Tsuboi, and K. Hirayama. 1993. Multifocal conduction block in n-hexane neuropathy. Muscle Nerve 16: 1416–1417.

Kuwabara, S. and K. Ogawara, J.Y. Sung, M. Mori, K. Kanai, T. Hattori, N. Yuki, C.S. Lin, D. Burke, and H. Bostock. 2002. Differences in membrane properties of axonal and demyelinating Guillain-Barré syndromes. Ann. Neurol. 52: 180–187.

Lacombblez, L. and G. Bensimon, P.N. Leigh, P. Guillet, and V. Meininger. 1996. Dose-ranging study of riluzole in amyotrophic lateral sclerosis. Amyotrofic lateral sclerosis/riluzole study group II. Lancet 347: 1425–1431.

Lai, E.C. and K.J. Felice, B.W. Festff, M.J. Gawel, D.F. Gelinas, R. Kratz, M.F. Murphy, H.M. Natter, F.H, Norris, and S.A. Rudnicki. 1997. Effect of recombinant human insulin-like growth factor-I on progression of ALS. A placebo-controlled study. The north America ALS/IGF-I study group. Neurology 49(6): 1621–1630.

Lasek, R.J. and H. Gainer, and J.L. Barker. 1977. Cell to cell transfer of glial proteins to the squid qiant axon: The glia neuron protein transfer hypothesis. J. Cell Biol. 74: 631–645.

Lasek, R.J. and J.A. Garner, and S.T. Brady. 1984. Axonal Transport of the Cytoplasmic Matrix. J. Cell Biol. 99(1): 212s–221o.

Lewis, R.A. and A J Sunner, M.J. Brown, and A.K. Asbury. 1982. Multifocal demyelinating neuropathy with persistent conduction block. Neurology 32: 958–964.

Lorente de Nó, R. 1947. A study of nerve physiology. Stud. Rockfeller Inst. Med. Res. Repr. 132: 1–548.

Lubetzki, C. and A. Williams, and B. Stakoff. 2005. Promotion repair in multiple sclerosis: problems and prospects. Curr. Opin. Neurol. 18: 237–244.

Ludolph, A.C. and T. Meyer, and M.W. Riepe. 1999. Antiglutamate therapy of ALS-which is the next step? J. Neurol. Sci. 55: 79–95.

Lupski, J.R., and R. Montes de Oca-Luna, S. Slaugenhaupt, L. Pentao, V. Guzzetta, B.J. Trask, O. Saucedo-Cardenas, D.F. Barker, J.M. Killian, C.A. Garcia, A. Chakravarti, and P.I. Patel. 1991. DNA duplication associated with Charcot-Marie-Tooth disease type 1A. Cell 66: 219–232.

Marques, W. Jr. and M.R. Freitas, O.J. Nascimento, A.B. Olivera, L. Calia, A. Melo, R. Lucena, V. Rocha, and A.A. Barreira. 2005. 17p duplicated Charcot-Marie-Tooth 1A: characteristics of a new population. J. Neurol. 252: 972–979.

Martini, R. 2001. The effect of myelinating Schwann cells on axon. Muscle Nerve 24: 456–466.

Martini, R. and K.V. Toyka. 2004. Immune-mediated components of hereditary demyelinating neuropathies: lessons from animal models and patients. Lancet Neurol. 3: 457–465.

McGregor, J.E. and Z. Wang, C. French-Constant, and A.M. Donald. Microscopy of mediation. pp. 1185–1195. In: A Mendez-Vilas and J. Diaz. [eds.]. 2010. Microscopy: Science, Technology, Applications and Education. Formatted.

Miler, J.H. and W.L. Rotten, and H.B. Boom. 1998. Extracellular potentials from active myelinated fibres inside insulated and no insulated peripheral nerves. IEEE Trans. Boomed. Eng. 45: 1146–1153.

Miller, R.G. and R. Sufi. 1997. New approaches to the treatment of ALS. Neurology 48(suppl. 4): S28–S32.

Moffett, J.R. and B. Ross, P. Arum, C.N. Madhavarao, and A.M. Namboodiri. 2007. N-Acetylaspartate in the CNS: from neurodiagnostics to neurobiology. Prog. Neurobiol. 81: 89–131.

Mogyoros, I. and M.C. Kiernan, D. Burke, and H. Bostock. 1998. Strength-duration properties of sensory and motor axons in amyotrophic lateral sclerosis. Brain 121: 851–859.

Moldovan, M. and S. Alvarez, and C. Krarup. 2009. Motor axon excitability during Wallerian degeneration. Brain 132: 511–523.

Moradmand, K. and M.D. Goldfinger. 1995. Computation of Long-Distance Propagation of Impulses Elicited by Poisson Process Stimulation. J. Neurophysiol. 74: 2415–2426.

Morland, C. and S. Henjum, E.G. Iversen, K.K. Skrede, and B. Hassel. 2007. Evidence for a higher glycolytic than oxidative metabolic activity in white matter of rat brain. Neurochem. Int. 50(5): 703–709.

Nakanishi, T. and M. Tamaki, and K. Arasaki. 1989. Maximal and minimal motor nerve conduction velocity in amyotrophic lateral sclerosis. Biology 39: 580–583.

Nave, K.A. 2010. Myelination and the trophic support of long axons. Nat. Rev. Neurosci. 11(4): 275–283.

Nave, K.A. and J.L. Salzer. 2006. Axonal regulation of myelination by neuregulin 1. Curr. Opin. Neurobiol. 16(5): 492–500.

Nave, K.A. and M.W. Sereda, and H. Ehrenreich. 2007. Mechanisms of disease: inherited demyelinating neuropathies—from basis to clinical research. Nat. Clin. Pract. Neurol. 3: 453–464.

Nave, K.A. and B.D. Trapp. 2008. Axon-Glial Signaling and the Glial Support of Axon Function. Annu. Rev. Neurosci. 31: 535–561.

Ndubaku, U. and M.E. de Bellard. 2008. Glial cells: Old cells with new twists. Acta Histochemica 110: 182–195.

Nernst, W. (1908). Zür theorie des elektrichen reizes. Pflügers Arch. 122: 275–314.

Nilsson, I. and C.H. Berthold 1988. Axon classes and internodal growth in the ventral spinal root L7 of adult and deviloping cats. J. Anat. 156: 71–96.

Nobile-Orazio, E. and N. Meucci, S. Barbieri, M. Carpo, and G. Scarlato. 1993. High-dose intravenous immunoglobulin therapy in multifocal motor neuropathy. Neurology 43: 537–544.

Nodera, H. and H. Bostock, S. Kuwabara, T. Sakamoto, K. Asanuma, J.Y. Sung, K. Ogawara, N. Hattori, M. Hirayama, G. Sobue, and R. Kaji. 2004. Nerve excitability properties in Charcot-Marie-Tooth disease type 1A. Brain 127: 203–211.

Nodera, H. and R. Kaji. 2006. Nerve excitability testing and its clinical application to neuromuscular diseases. Clin. Neuphysiol. 117: 1902–1916.

Ochoa, J. The unmyelinated nerve fibre. pp. 106–158. *In*: D.N. Landon [ed.]. 1976. The peripheral nerve. Champan and Hill, London.

Oguivetskaia, K. and C. Cifuentes-Diaz, J.A. Girault, and L. Goutebroze. 2005. Cellular contacts in myelinated fibers of the peripheral nervous system. Med. Sci. 21: 162–169.

Ohnishi, A. and A. Mitsudome, and Y. Muray. 1987. Primary segmental demyelination in the sural nerve in Cockayne's syndrome. Muscle Nerve 10: 163–167.

Panizza, M. and J. Nilsson, B.J. Roth, J. Rothwell, and M. Hallett. 1994. The time constants of motor and sensory peripheral nerve fibers measured with the method of latent addition, Electroencephalography and Clin. Neurophysiol. 93: 147–154.

Panizza, M. and J. Nilsson, B.J. Roth, S.E. Grill, M. Demirsci, and M. Hallett. 1998. Differences between the strength-duration time constants of sensory and motor peripheral nerve fiber: further studies and conditions. Muscle Nerve 21: 48–54.

Parry, G.J. 1985. Are multifocal motor neuropathy and Lewis-Sumner syndrome distinct nosologic entities? Muscle Nerve 22: 557–559.

Parry, G.J. Multifocal motor neuropathy: Pathology and treatment. p. 73. *In*: J. Kimura and R. Kaji. [eds.]. 1997. Physiology of ALS and related diseases. Elsevier Science BV, Amsterdam.

Parry, G.J. and S. Clarke. 1985. Pure motor neuropathy with multifocal conduction block masquerading as motor neuron disease. Muscle Nerve 8: 617–619.

Pedraza, L. and J.K. Huang, and D.R. Colman. 2001. Organizing principles of the axoglial apparatus. Neuron 30: 335–344.

Peles, E. and J.L. Salzer. 2000. Molecular domains of myelinated axons. Review. Curr. Opin. Neurobiol. 10: 558–565.

Pestronk, A. and D.R. Cornblath, A.A. Ilyae, H. Baba, R.H. Quarles, J.W. Griffin, K. Alferson, and A. Adams. 1988. A treatable multifocal motor neuropathy with antibodies to GM1 ganglioside. Ann. Neurol. 24: 73–78.

Plonsey, R. 1977. Action potential sources and their volume conducted fieldes. Proc. IEEE 65: 601–609.

Pollard, J.D. and J.G. McLeod, P. Gatenby, and H. Kronenberg. 1983. Prediction of response to plasma exchange in chronic relapsing polyneuropathy. J. Neurol. Sci. 58: 269–287.

Priori, A. and B. Bossi, G. Ardolino, L. Bertolasi, M. Carpo, E. Nobile-Orazio, and S.Barbieri. 2005. Pathophysiological heterogeneity of conduction blocks in multifocal motor neuropathy. Brain 128: 1642–1648.

Quandt, F.N. and F.A. Davis. 1992. Action potential refractory period in axonal demyelination: computer simulation. Biol. Cybern. 67: 545–552.

Quarles, R.H. and W.B. Macklin, and P. Morell. Myelin formation, structure and biochemistry. *In:* G.H. Siegel, B.W. Agranoff, R.W. Albers, S.K. Fisher, and M.D. Uhler. [eds.]. 2006. Basic Neurochemistry: Molecular, Cellular and Medical Aspects, Elsevier, Inc.

Raeymaekers, P. and V. Timmerman, E. Nelis, P. De Jonghe, J.K. Hoogendijk, F. Bass, D.F. Barker, J.J. Martin, M. De Visser, P.A. Bolhuis, and C. Van Broeckhover. 1991. HMSN Collaborative Study Group, Duplication in chromosome 17p11.2 in Charcot-Marie-Tooth disease type 1A (CMT1A). Neuromuscul. Disord. 1: 93–97.

Rasband, M.N. 2011. Composition, assembly, and maintenance of excitable membrane domains in myelinated axons. Seminars in Cell and Developmental Biology 22: 178–184.

Rasband, M.N. and J.S. Trimmer, T.L. Schwarz, S.R. Levinson, M.N Ellisman, M. Scharchner, and P. Shrager. 1998. Potassium channel distribution, clustering, and function in remyelinating rat axons. J. Neurosci. 18: 36–47.

Rasband, M.N. and E. Peles, J.S. Trimmer, S.R. Levinson, S.E. Lux, and P. Shrager. 1999. Dependence of nodal sodium channels clustering on paranodal axoglial contact in the developing CNS. J. Neurosci. 19(17): 7516–7528.

Rezania, K. and B. Gundogdu, and B. Soliven. 2004. Pathogenesis of chronic inflammatory demyelinating polyradiculoneuropathy. Front Biosci. 9: 939–945.

Rinzel, J. A formal classification of bursting mechanisms in excitable systems. pp. 1578–1594. *In:* A.M. Gleason [ed.]. 1987. Proceedings of the International Congress of Mathematics. American Mathematical Society, Providence.

Rios, J.C. and C.V. Melendez-Vasquez, S. Einheber, M. Lustig, M. Grumet, J. Hemperly, E. Peles, and J.L. Salzer. 2000. Contractin-associated protein (Caspr) and contractin form a complex that is targeted to the paranodal junctions during myelination. J. Neurosci. 20(22): 8354–8364.

Rios, J.C. and M. Rubin, M. St Martin, R.T. Downey, S. Einheber, J. Rosenbluth, S.R. Levinson, M. Bhat, and J.L. Salzer. 2003. Paranodal interactions regulate expression of sodium channel subtypes and provide a diffusion barrier for the node of Ranvier. J. Neurosci. 23(18): 7001–7011.

Risling, M. and C. Hildebrand. 1982. Occurrence of unmyelinated axon profiles at distal middle and proximal levels in the ventral root L7 of cats and kittens. J. Neurol. Sci. 56: 219–231.

Ritchie, J.M. and S.Y. Chiu. Distribution of sodium and potassium channels in mammalian myelinated nerve. pp. 329–342. In: S.G. Waxman and J.M. Ritchie [eds.]. 1981. Demyelinating diseases: basic and clinical electrophysiology. Raven Press, New York.

Roomi, M.W. and A. Ishaque, N.R. Khan, and E.H. Eylar. 1978. The P0 protein. The major glycoprotein of peripheral nerve myelin. Biochim. Biophys. Acta 536: 112–121.

Roper, J. and J.R. Schwarz. 1989. Heterogeneous distribution of fast and slow potassium channels in myelinated rat nerve fibers. J. Physiol. (Lond.) 416: 93–110.

Ropte, S. and P. Scheidt, and R.L. Friede. 1990. The intermediate dense line of the myelin sheath is preferentially accessible to cations and is stabilized by cations. J. Neurocitol. 19: 242–252.

Rosen, D.R., T. Siddigue, D. Patterson, D.A. Figlewicz, P. Sapp, D. Hentati, D. Donaldson, J. Goto, J.P. O'Regan, H.X. Deng, et al. 1993. Mutations in Cu/Zn superoxide dismutase gene are associated with familial amyotrophic lateral sclerosis. Nature 362(6415): 59–62.

Rosenblueth, A. 1941. The effects of direct currents on the electrical excitability of nerve. Am. J. Physiol. 132: 53–73.

Rosenfalck, P. 1969. Intra- and extracellular potential fields of active nerve and muscle fibres. A physicomathematical analysis of different models. Thromb Diath Haemorrh Suppl. 321: 1–168.

Roth, G. 1982. The origin of fasciculations. Ann. Neurol. 12: 542–547.

Roth, G. 1984. Fasciculations and their F-response. Localization of their axonal origin. J. Neurol. Sci. 63: 299–306.

Rowinska-Marcinska, K. and B. Ryniewicz, I. Hansmanowa-Petrusewicz, and K. Karvanska. 1997. Diagnostic values of satellite potentials in clinical EMG. Electromyogr. Clin. Neurophysiol. 37: 483–489.

Sabatelli, M., T. Mignogna, G. Lippi, et al. 1995. Interferon-α may benefit steroid unresponsive chronic inflammatory demyelinating polyneuropathy. J. Neurol. Neurosurg. Psychiatry 58: 321–328.

Safronov, B.V. and K. Kampe, and W. Vogel. 1993. Single voltage-dependent potassium channels in rat peripheral nerve membrane. J. Physiol. (Lond.) 460: 675–691.

Sahenk, Z. and H.N. Nagaraja, B.S. McCracken, W.M. King, M.L. Freimer, J.M. Cedarbaum, and J.R. Mendell. 2005. NT-3 promotes nerve regeneration and sensory improvement in CMT1A mouse models and in patients. Neurology 65(5): 681–689.

Saher, G. and S. Quintes, and K.A. Nave. 2011. Cholesterol: A novel regulatory role in myelin formation. Neuroscientist 17(1): 79–93.

Salzer, J.L. 1997. Clustering Sodium Channels at the Node of Ranvier: Close Encounters of the Axon-Glia Kind. Neuron 18: 843–846.

Salzer, J.L. 2003. Polarized domains of myelinated axons. Neuron 38: 297–318.

Salzer, J.L. and P.L. Brophy, and E. Peles. 2008. Molecular domains of myelinated axons in the peripherial nervous system. Glia 56: 1532–1540.

Saperstein, D.S. and J.S. Katz, A.A. Amato, and R.J. Barohn. 2001. Clinical spectrum of chronic acquired demyelinating polyneuropathy. Muscle Nerve 24(3): 311–324.

Schauf, C.L. and F.A. Davis. 1974. Impulse conduction in multiple sclerosis: a theoretical basis for modification by temperature and pharmacological agents. J. Neurol. Neurosurg. Psychiatry 37: 152–161.

Schiefer, M.A. 2009. Optimized design of neural interfaces for femoral nerve clinical neuroprostheses: Anatomically-based modeling and intraoperative evaluation. Ph.D. Thesis, Department of Biomedical Engineering, Case Western Reserve University.

Schmidt, B. and G. Stoll, P. Van Der Meide, S. Jung, and H.P. Hartung. 1992. Transient cellular expression of γ-interferon in myelin-induced and T-cell line-mediated experimental autoimmune neuritis. Brain 115: 1633–1646.

Scholz, A. and G. Reid, W. Vogel, and H. Bostock. 1993. Ion channels in human axons. J. Neurophysiol. 70: 1274–1279.

Schwarz, J.R. and G. Reid, and H. Bostock. 1995. Action potentials and membrane currents in the human node of Ranvier. Pflügers Arch. 430: 283–292.

Sheikh, K.A. and T.D. Deerinck, M.H. Ellisman, and J.W. Griffin. 1999. The distribution of ganglioside-like moieties in peripheral nerves. Brain 122(3): 449–460.

Sherman, D.L. and P.J. Brophy. 2005. Mechanisms of axon ensheathment and myelin growth. Review. Nat. Rev. Neurosci. 6(9): 683–690.

Sherman, D.L. and S. Tait, S. Melrose, R. Johnson, B. Zonta, F.A. Court, W.B. Macklin, S. Meek, A.J. Smith, D.F. Cottrell, and P.J. Brophy. 2005. Neurofascins are required to establish axonal domains for saltatory conduction. Neuron. 48(5): 737–742.

Shrager, P. 1989. Sodium channels in single demyelinated mammalian axons. Brain Res. 483: 149–154.

Shy, M.E. 2006. Peripheral neuropathies caused by mutations in the myelin protein zero. J. Neurol. Sci. 242: 55–66.

Shy, M.E. and J.Y. Garben, and J. Kamholz. 2002. Hereditary motor and sensory neuropathies: a biological perspective. Lancet Neurol. 1: 110–118.

Simons, M. and J. Trotter. 2007. Wrapping it up: the cell biology of myelination. Curr. Opin. Neurobiol. 17: 533–540.

Smith, R.S. and Z.J. Koles. 1970. Myelinated nerve fibers: Computed effect of myelin thickness conduction velocity. Am. J. Physiol. 219: 1256–1258.

Sobue, G. and Y. Hashizume, K. Sahashi, A. Takahashi, E. Mukai, Y. Matsuoka, and M. Mukoyama. 1983. Amyotrophic lateral sclerosis. Lack of central chromatolytic response of motor neurocytons corresponding to active axonal degeneration. Arch. Neurol. 40: 306–309.

Stampfly, R. 1959. Is the resting potential of Ranvier nodes a potassium potential? Ann. N.Y. Acad. Sci. 81: 265–284.

Stegeman, D.F. and J.P. de Weerd, and E.G. Eijkman. 1979. A volume conductor study of compound action potentials of nerves *in situ*: the forword problem. Biol. Cybern. 33: 97–111.

Stephanova, D.I. 1988a. Systematic paranodal demyelination of nerve fibres: computer simulations. Electromyogr. Clin. Neurophysiol. 28: 107–110.

Stephanova, D.I. 1988b. Reorganization of the axonal membrane in a demyelinated nerve fibre: computer simulation. Electromyogr. Clin. Neurophysiol. 28: 101–105.

Stephanova, D.I. 1988c. The effect of temperature on a simulated systematically paranodally demyelinated nerve fibre. Biol. Cybern. 60: 73–77.

Stephanova, D.I. 1989a. Model investigations of the temperature dependence of demyelinated and reorganized axonal membrane. Biol. Cybern. 60: 439–443.

Stephanova, D.I. 1989b. Conduction along myelinated and demyelinated nerve fibres during the recovery cycle: model investigation. Biol. Cybern. 62: 83–87.

Stephanova, D.I. 1990. Conduction along myelinated and demyelinated nerve fibres with a reorganized axonal membrane during the recovery cycle: model investigations. Biol. Cybern. 64: 129–134.

Stephanova, D.I. 2001. Myelin as longitudinal conductor: a multi-layered model of the myelinated human motor nerve fibre. Biol. Cybern. 84: 301–308.

Stephanova, D.I. 2006. Excitability and potentials of human fibres in amyotrophic lateral sclerosis: model investigations. pp. 155–178. *In:* C.A. Murray [ed.]. Amyotrophic Lateral Sclerosis: New Research. Nova Science Publishers Inc., New York.

Stephanova, D. 2010. Comparison of membrane property abnormalities in simulated demyelinating neuropathies and neuronopathies. Compt. rend. Acad. bulg. Sci. 63(3): 435–442.

Stephanova, D.I. and A.S. Alexandrov. 2006. Simulating mild systematic and focal demyelinating neuropathies: membrane property abnormalities. J. Integr. Neurosci. 5: 595–623.

Stephanova, D.I. and A.S. Alexandrov, A. Kossev, and L. Christova. 2007a. Simulating focal demyelinating neuropathies: membrane property abnormalities. Biol. Cybern. 96: 195–208.

Stephanova, D.I. and H. Bostock. 1995. A distributed-parameter model of the myelinated human motor nerve fibre: temporal and spatial distributions of action potentials and ionic currents. Biol. Cybern. 73: 275–280.

Stephanova, D.I. and H. Bostock. 1996. A distributed-parameter model of myelinated human motor nerve fibre: temporal and spatial distributions of electrotonic potentials and ionic currents. Biol. Cybern. 74: 543–547.

Stephanova, D.I. and M. Daskalova. 2002. Extracellular potentials of human motor myelinated nerve fibres in normal case and in amyotrophic lateral sclerosis. Electromyogr. Clin. Neurophysiol. 42: 443–448.

Stephanova, D.I. and M. Daskalova. 2005a. Differences in potentials and excitability properties in simulated cases of demyelinating neuropathies. Part II. Paranodal demyelination. Clin. Neurophysiol. 116: 1159–1166.

Stephanova, D.I. and M. Daskalova. 2005b. Differences in potentials and excitability properties in simulated cases of demyelinating neuropathies. Part III. Paranodal internodal demyelination. Clin. Neurophysiol. 116: 2334–2341.

Stephanova, D.I. and M. Daskalova. 2008. Membrane property abnormalities in simulated cases of mild systematic and severe focal demyelinating neuropathies. Eur. Biophys. J. 37: 183–195.

Stephanova, D.I. and M. Daskalova. 2009. Homogeneity or heterogeneity membrane polarization in simulated cases of demyelinating neuropathies without or with conduction block. Compt. Rend. Acad. Bulg. Sci. 62(8). 1015–1022.

Stephanova, D.I. and M. Daskalova, and A.S. Alexandrov. 2005. Differences in potentials and excitability properties in simulated cases of demyelinating neuropathies. Part I. Clin. Neurophysiol. 116: 1153–1158.

Stephanova, D.I. and M. Daskalova, and A.S. Alexandrov. 2006a. Differences in membrane properties in simulated cases of demyelinating neuropathies. Internodal focal demyelination without conduction block. Journal of Biological Physics 32: 61–71.

Stephanova, D.I. and M. Daskalova, and A.S. Alexandrov. 2006b. Differences in membrane properties in simulated cases of demyelinating neuropathies. Internodal focal demyelination with conduction block. Journal of Biological Physics 32: 129–144.

Stephanova, D.I. and M. Daskalova, and A.S. Alexandrov. 2007b. Channels, currents and mechanisms of accommodative processes in simulated cases of systematic demyelinating neuropathies. Brain Research 1171: 138–151.

Stephanova, D.I. and S.M. Krustev, and M. Daskalova M. 2011a. The aqueous layers within the myelin sheath modulate the membrane properties of simulated hereditary demyelinating neuropathies, J. Integr. Neurosci. 10(1): 89–103.

Stephanova, D.I. and S.M. Krustev, N. Negrev, and M. Daskalova. 2011b. The myelin sheath aqueous layers improve the membrane properties of simulated chronic demyelinating neuropathies, J. Integr. Neurosci. 10(1): 105–120.

Stephanova, D.I. and S.M. Krustev, and N. Negrev. 2012a. Mechanisms defining the action potential abnormalities in simulated amyotriphic lateral sclerosis. J. Integr. Neurosci. 11(2): 137–154.

Stephanova, D.I. and S.M. Krustev, and N. Negrev. 2012b. Mechanisms defining the electrotonic potential abnormalities in simulated amyotriphic lateral sclerosis. J. Integr. Neurosci. 11(2): 155–167.

Stephanova, D.I. and K. Mileva. 2000. Different effects of blocked potassium channels on action potentials, accommodations, adaptation and anode break excitation in human motor and sensory myelinated nerve fibers: computer simulations. Biol. Cybern. 83: 161–167.

Stephanova, D.I. and N. Trayanova, A. Gydikov, and A. Kossev. 1989. Extracellular potentials of a single myelinated nerve fiber in an unbounded volume conductor. Biol. Cybern. 61: 205–210.

Story, J.S. and L.H. Phillips. 1995. A clinical approach to the patient with chronic motor neuropathy. Neurologist 1: 134–145.

Strujik, J.J. 1997. The extracellular potential of a myelinated nerve fiber in an unbounded medium and in nerve cuff models. Biophys. J. 72: 2457–2469.

Sumner A.J. Consensus criteria for the diagnosis of partial conduction block and multifocal motor neuropathy. pp. 221–232. *In:* J. Kimura and R. Kaji [eds.]. 1997. Physiology of ALS and related Diseases, Elsevier Science BV, Amsterdam.

Sung, J.Y. and S. Kuwabara, R. Kaji, K. Ogawara, M. Mori, K. Kanai, H. Nodera, T. Hattori, and H. Bostock. 2004. Threshold electrotonus in chronic inflammatory demyelinating polyneuropathy: correlation with clinical profiles. Muscle Nerve 29: 28–37.

Suter, U. and S.S. Scherer. 2003. Disease Mechanisms in inherited neuropathies. Nature Reviews Neuroscience 4: 714–726.

Tan E. and M. Hajinazarian, W. Baw, J. Neff, and J.R. Mendell. 1993. Acute renal failure resulting from intravenous immunoglobulin therapy. Arch. Neurol. 50: 137–139.

Tasaki, I. 1955. New measurements of the capacity and the resistance of the myelin sheath and the nodal membrane of the isolated frog nerve fiber. Am. J. Physiol. 181: 639–650.

Thaxton, C. and M. Bhat. 2009. Myelination and regional domain differentiation of the axon. Results Probl. Cell Differ. 48: 1–28.

Toyka, K.V. and R. Augspach, W. Paulus, B. Grabensec, and D. Hein. 1980. Plasma exchange in polyradiculoneuropathy. Ann. Neurol. 8: 205–206.

Trapp, B.D. and G.J. Kidd. Structure of myelinated axon. pp.3–27. *In:* R.A. Lazzarini, J.W. Griffin, H. Lassman, K.A. Nave, R.H. Miller, and B.D. Trapp [eds.]. 2004. Myelin biology and disorders. Elsevier Academic Press, San Diego.

Trapp, B.D. and K.A. Nave. 2008. Multiple sclerosis: an immune or neurodegenerative disorder? Annu. Rev. Neurosci. 31: 247–269.

Trayanova, N.A. and C.S. Henriquer, and R. Plonsey. 1990. Limitations of approximate solution for computing the extracellular potentials of single fibers and bundle equations. IEEE Trans. Biomed. Eng. 39: 22–35.

Trojaborg, W. and F. Buchthal. 1965. Malignant and bening fasciculations. Acta Neurol. Scand. Suppl. 13: 251–254.

van der Bergh, P. and E.L. Logigian, and J.J. Kelly. 1989. Motor neuropathy with multifocal conduction blocks. Muscle Nerve 11: 26–31.

van Doorn, P.A. and M. Verneulen, A. Brand, P.G.H. Mulder, and H.F.M. Busch. 1991. Intravenous immunoglobulin treatment in patients with chronic inflammatory demyelinating polyneuropathy. Arch. Neurol. 48: 217–220.

Verhamme, C. and I.N. van Schaik, J.H. Koelman, R.J. de Haan, and M. de Visser. 2009. The natural history of Charcot-Marie-Tooth type 1A in adults: a 5-year follow-up study. Brain 132(12): 3252–3262.

Vermeulen M. and P.A. van Doorn, A. Brand, P.F.W. Strengers, F.G.I. Jennekens, and H.F.M. Busch. 1993. Intravenous immunoglobulin treatment in patients with chronic inflammatory demyelinating polyneuropathy: Double blind placebo controlled study. J. Neurol. Neurosurg. Psychiatry 56: 36–39.

Vogel, W. and J.R. Schwarz. Voltage-clamp studies in axons: macroscopic and single-channel currents. pp. 257–280. *In:* S.G. Waxman, J.D. Kocsis, and P.K. Stys [eds.]. 1995. The axon, Oxford University Press, New York.

Vriesendorp, F.J. and R.F. Mayer, and C.L. Koski. 1991. Kinetics of anti-peripheral nerve myelin antibody in patients with Guillain Barré syndrome treated and not treated with plasmapheresis. Arch. Neurol. 48: 858–861.

Wang, X.J. and J. Rinzel. Oscillatory and bursting properties of neurons. pp. 686–691. *In:* M. Arbib [ed.]. 1995. The handbook of brain theory and neural network. MIT Press, Cambridge Mass.

Waxman, S.G. 1977. Conduction in myelinated, unmyelinated, and demyelinated fibers. Arch. Neurol. 34: 585–590.

Waxman, S.G. 1978. Prerequisites for conduction in demyelinated fibres. Neurology 28: 27–33.

Waxman, S.G. 1997. Axon-glia interactions: Building a smart nerve fiber. Current Biology 7: R406–R410.

Waxman, S.G. and M.H. Brill. 1978. Conduction through demyelinated plaques in multiple sclerosis: Computer simulations of facilitation by short internodes. J. Neurol. Neurosurg. Psychiatry 41: 408–416.

Waxman, S.G. and J.D. Kocsis, M.H. Brill, and H.A. Swadlow. 1979. Dependence of refractory period measurements on conduction distance: a computer simulation analysis. Electroencephalogr. Clin. Neurophysiol. 47: 717–724.

Waxman, S.G. and J.M. Ritchie. 1993. Molecular dissection of the myelinated axon. Ann. Neurol. 33: 121–136.

Waxman, S.G. and S.L. Wood. 1984. Impulse conduction in inhomogeneous axons: effect of variation of voltage-sensitive conductances on invasion of demyelinated axon segments and preterminal fibers. Brain Res. 294: 111–122.

Weiss, G. 1901. Sur la possibilité de rendre comparables entre eux les appareils servant à l'excitation électrique. Arch. Ital. Biol. 35: 413–446.

Wettstein, A. 1979. The origin of fasciculations in motoneuron disease. Ann. Neurol. 5: 295–300.

Willians, K.A. and C.M. Deber. 1993. The structure and function of central nervous system myelin. Crit. Rev. Clin. Lab. Sci. 30: 29–64.

Winckler, B. and P. Forscher, and I. Mellman. 1999. A diffusion barrier maintains distribution of membrane proteins in polarized neurons. Nature 397: 698–701.

Wirguin, I. and T. Brenner, Z. Argov, and I. Steiner. 1992. Multifocal motor nerve conduction abnormalities in amyotrophic lateral sclerosis. J. Neurol. Sci. 112: 199–203.

Wood, S.L. and S.G. Waxman. 1982. Conduction in demyelinated nerve fibers: Computer simulations of the effects of variation in voltage sensitive conductances. Proc. IV Ann. IEEE Frontiers of Engineering in Health Care 11.7.1: 424.

Wood, S.L. and S.G. Waxman, and J. Kocsis. 1982. Conduction of trains of stimuli in uniform myelinated fibers: Computer dependence on stimulus frequency. Neurosci. 7(2): 423–430.

Yates, A.J. and J.P. Bouchard, and J.T. Wherrett. 1976. Relation of axonal membrane to myelin membrane in sciatic nerve during development: Comparison of morphological and chemical parameters. Brain Res. 104: 261–271.

Yim, S.Y. and I.Y. Lee, H.W. Moon, U.W. Rah, S.H. Kim, C. Shim, and H.J. Joo. 1995. Hypertrophic Neuropathy with Complete Conduction Block–Hereditary motor and sensory neuropathy type III–, Yonsei Medical Journal 36(5): 466–472.

Yiu, E.M. and J. Burns, M.M. Ryan, and R.A. Ouvrier. 2008. Neurophysiologic abnormalities in children with Charcot-Marie-Tooth disease type 1A. J. Peripher. Nerv. Syst. 3(3): 236–241.

Yokota, T. and Y. Salto, N. Yuki, and H. Tanaka. 1996. Persistent increased threshold of electrical stimulation selective to motor nerve in multifocal motor neuropathy. Muscle Nerve 19: 823–828.

Young, P. and U. Suter. 2001. Disease mechanisms and potential therapeutic strategies in Charcot-Marie-tooth disease. Brain Res. Rev. 36: 213–221.

Young, P. and U. Suter. 2003. The causes of Charcot-Marie-tooth disease. Cell Mol. Life 60: 2547–2560.

Zhou, L. and C.L. Zhang, A. Messing, and S.Y. Chiu. 1998. Temperature-sensitive neuromuscular transmission in Kv1.1 null mice: role of potassium channels under the myelin sheath in young nerves. J. Neurosci. 1: 7200–7215.

Zlochiver, S. 2010. Persistent reflection underlies ectopic activity in multiple sclerosis: a numerical study. Biol. Cybern. 102(3): 181–196.

Zu Hörste, G.M. and K.A. Nave. 2006. Animal models of inherited neuropathies. Curr. Opin. Neurol. 19(5): 464–479.

Index